T0269644

Animal Vigilance

Animal Vigilance
Monitoring Predators and Competitors

Guy Beauchamp

ELSEVIER

AMSTERDAM • BOSTON • HEIDELBERG • LONDON
NEW YORK • OXFORD • PARIS • SAN DIEGO
SAN FRANCISCO • SINGAPORE • SYDNEY • TOKYO
Academic Press is an Imprint of Elsevier

Academic Press is an imprint of Elsevier
125, London Wall, EC2Y 5AS, UK
525 B Street, Suite 1800, San Diego, CA 92101-4495, USA
225 Wyman Street, Waltham, MA 02451, USA
The Boulevard, Langford Lane, Kidlington, Oxford OX5 1GB, UK

Notices
Knowledge and best practice in this field are constantly changing. As new research and experience
broaden our understanding, changes in research methods, professional practices, or medical treatment may become necessary.

Practitioners and researchers must always rely on their own experience and knowledge in evaluating
and using any information, methods, compounds, or experiments described herein. In using such
information or methods they should be mindful of their own safety and the safety of others, including parties for whom they have a professional responsibility.

To the fullest extent of the law, neither the Publisher nor the authors, contributors, or editors, assume
any liability for any injury and/or damage to persons or property as a matter of products liability,
negligence or otherwise, or from any use or operation of any methods, products, instructions, or
ideas contained in the material herein.

British Library Cataloguing-in-Publication Data
A catalogue record for this book is available from the British Library

Library of Congress Cataloging-in-Publication Data
A catalog record for this book is available from the Library of Congress

ISBN: 978-0-12-801983-2

For information on all Academic Press publications
visit our website at http://store.elsevier.com/

Publisher: Janice Audet
Acquisition Editor: Kristi A.S. Gomez
Editorial Project Manager: Pat Gonzalez
Production Project Manager: Caroline Johnson
Designer: Ines Cruz

Typeset by Thomson Digital

Printed and bound in the United States of America

Working together
to grow libraries in
developing countries

www.elsevier.com • www.bookaid.org

Contents

Preface

Wildlife enthusiasts all know firsthand how difficult it is to approach animals undetected. Fleeing individuals are too often all we can glimpse even after a careful approach. Vulnerable animals possess an uncanny ability to detect threats from afar. The reason for this is that animals are ever vigilant for any cue in their surroundings that signals danger. There must be strong selection pressure to maintain a vigilant state. Nonchalant prey animals, for instance, would detect predators too late to escape successfully. Similarly, indifferent group members would miss cues that rivals covet their resources. Vigilance can increase the ability to detect and assess such threats, but it comes at the cost of diverting attention from other fitness-enhancing activities such as foraging and resting. Animals must therefore balance vigilance and competing activities to increase their chances of surviving or deterring attacks while avoiding starvation.

The idea that vigilance plays a role in reducing predation pressure was first formulated in the scientific literature more than 100 years ago by British naturalists like Galton and Bates. Modern vigilance research started with the formulation of a mathematical model by Pulliam in 1973, which describes how scanning the surroundings can increase predator detection. The model showed, not surprisingly, that detection should be more likely when the predator takes longer to mount an attack. In addition, when a predator targets individuals in a group, detection should occur earlier because of the multiplicity of the senses commanding all approaches. More surprisingly, extra safety in a group could allow individuals to reduce their own vigilance at no increased risk to themselves. In the following decades, animal vigilance students tested these predictions in a wide range of species and also uncovered many internal or external factors driving vigilance levels. These different factors affect the risk perceived by animals and thus modulate the expression of vigilance.

Anti-predator vigilance is but one tactic deployed by prey animals to avoid predation (Caro, 2005). Animals can reduce predation risk by hiding from predators. This can be achieved by camouflage or spatial aggregation (Ruxton et al., 2004). After an encounter with a predator, prey animals can reduce the risk of capture by adopting defensive formations or by using chemical warfare. Anti-predator vigilance fits in the middle of the predation sequence by providing a means to detect predators early and initiate escape more rapidly. Animals that rely on camouflage or active deterrence of predators probably benefit little from vigilance because they do not need to escape rapidly. For the others, vigilance is a tactic of choice to reduce predation pressure.

While much of the attention tended to focus on the relationship between vigilance and predation risk, it has become abundantly clear over the years that animals pay attention not only to predators but also to their neighbours. Nearby individuals compete for resources like food or mates. Vigilance can help detect and avoid aggressive neighbours. Vigilance is now considered a general avoidance tactic aimed not only at predators, but also at rivals and foes from the same species.

Animal vigilance has been at the forefront of research on animal behaviour for the past 40 years, and yet there has been no comprehensive review of this topic. Students of animal behaviour have focused on many aspects of animal vigilance, from models of its adaptive value to empirical research in the laboratory and in the field. The vast literature on vigilance is widely dispersed with often little contact between theoretical and empirical researchers and between researchers focusing on different taxa like birds and mammals. In this book, I aim to provide such a comprehensive review by considering vigilance from all angles, theoretical and empirical, while including the broadest range of species to underscore unifying themes.

Chapter 1 presents an overview of this field of research and tackles the tricky definition of vigilance in animals. I also discuss the various types of vigilance that exist in animals, including a distinction between reactive and proactive vigilance and between vigilance aimed at predators (anti-predator vigilance) and competitors (social vigilance). Using text mining of vigilance references, I present research themes that have emerged over the years. I also provide a brief history of vigilance research from its beginnings 100 years ago to the present.

In Chapter 2, I focus on the function of vigilance during both the breeding and non-breeding seasons. Vigilance can be used to monitor predators as well as rivals and foes competing for resources such as food and mates. The functional approach has dominated the field of animal vigilance research. Nevertheless, developmental, mechanistic and evolutionary issues are also important to provide complementary explanations of vigilance. These issues have been unjustly neglected. I cover them in Chapter 3 and highlight several emerging trends worthy of future research.

In Chapter 4, I provide a broad discussion on internal and external factors that modulate the expression of vigilance. Internal factors include reserve levels, for example, and external factors involve environmental features such as distance to cover, weather or inter-individual distances. I refer to these factors as drivers of vigilance; they influence the social or predation risk experienced by individuals and as such they have predictable effects on vigilance. For example, prey species that rely on cover to escape from predators are expected to increase their vigilance when foraging farther from protective cover. I explain how each driver is expected to affect vigilance and examine the evidence for and against each hypothesis.

Group size is the driver of vigilance that has received the most attention in the literature. I devote two full chapters to the effect of group size on vigilance. In Chapter 5, I present the theory that explains why vigilance is expected to decline with group size. Various mechanisms act together to modulate vigilance in a group including the many-eyes effect, collective detection and risk dilution (Lima and Dill, 1990). All these issues are explored in non-mathematical terms. I restrict mathematical details to separate boxes in the text. Models make necessary but often debatable assumptions about the vigilance process. In particular, the issue of randomness in the temporal organization of vigilance bouts has attracted much attention. I review the evidence supporting (or not) these assumptions later in the chapter.

In Chapter 6, I review the vast literature on the effect of group size on vigilance in birds and in mammals. I examine the reasons why vigilance may not always be expected to decrease with group size. I then examine possible interactions between environmental factors and group size, which can explain why the magnitude of the group-size effect on vigilance often varies within the same species. There are alternative explanations as to why vigilance should decrease with group size. In particular, I discuss how competition for resources can lead to a decrease in vigilance without invoking predation risk.

Several assumptions used in influential models of vigilance have been challenged recently. The assumption that vigilance is independent amongst group members, in particular, has been tested and found wanting in many species. Individuals often copy the vigilance state of their neighbours, and vigilance in a group tends to be synchronized rather than independent. I explain why synchronization makes evolutionary sense and review the available evidence in Chapter 7. Co-ordination of vigilance by different individuals to ensure a more constant level of vigilance in the group represents yet another example of non-independent vigilance. This is particularly well documented in species with sentinels. I review models and the evidence for co-ordination of vigilance later in the same chapter.

In Chapter 8, I explore vigilance in contexts that have received less attention. For instance, what happens to vigilance in species or in habitats where predation risk, one of the main drivers of vigilance, is relaxed? In addition, what happens to vigilance in groups that include more than one species? Exploring vigilance outside the box is certainly a way to maintain this topic at the forefront of animal behaviour research. In the conclusion, I highlight promising avenues for future research on animal vigilance.

The literature on animal vigilance is vast and while I tried to be thorough in citing relevant papers, I may have missed some references. This in no way reflects the value of these papers. I have been working on vigilance projects for nearly 20 years now and benefited greatly from the help and expertise of many collaborators including Peter Alexander, Peter Bednekoff, Dan Blumstein, Bill Cooper, Robin Dunbar, Esteban Fernández-Juricic, Luc-Alain Giraldeau, Eben Goodale, Andrew Jackson, Roger Jovani, Chunlin Li, Zhongqiu Li, Barbara

Livoreil, Raymond McNeil, Olivier Pays, Graeme Ruxton, Étienne Sirot, Jian-bin Shi and Hari Sridhar. I am also thankful for many insightful exchanges with John Lazarus about the early days of vigilance research. The staff at Academic Press was ever helpful, and it is a pleasure to thank Kristi Gomez and Patricia Gonzalez for getting the project safely home.

Chapter 1

Overview of Animal Vigilance

1.1 INTRODUCTION

Victorian England produced its fair share of famous polymaths. Chief amongst them is Francis Galton (1822–1911) who made seminal contributions to fields of research as varied as psychology, meteorology, and genetics. Better known for the development of eugenics, the improvement of the human race by selective breeding, he also made several long-standing contributions in less controversial fields (Brookes, 2003). Buried in his massive output, there is a record of a trip to present-day Namibia during which he observed the behaviour of free-ranging Damara cattle. While he was primarily interested in understanding the slavish attitudes of men, which we know today as instincts, he was also drawn to social animals, such as the ox, to better understand the gregarious instinct (Galton, 1883). At the time of his research, African lions often ambushed grazing Damara cattle. Galton made the following observations on the cattle:

> When he is alone it is not simply that he is too defenceless, but that he is easily surprised. ...cattle are obliged in their ordinary course of life to spend a considerable part of the day with their heads buried in the grass, where they can neither see nor smell what is about them. ...But a herd of such animals, when considered as a whole, is always on the alert; at almost every moment some eyes, ears, and noses will command all approaches, and the start or cry of alarm of a single beast is a signal to all his companions. ...The protective senses of each individual who chooses to live in companionship are multiplied by a large factor, and he thereby receives a maximum of security at a minimum cost of restlessness.

As would be expected from a cousin of Charles Darwin, the father of natural selection, Galton later concluded that there is little doubt that gregariousness in cattle evolved to reduce predation risk.

This excerpt clearly illustrates several key concepts in the study of animal vigilance. Animals use various senses to monitor their surroundings for potential threats, such as ambushing lions in the case of cattle. The purpose of such monitoring is the early detection of threats. Upon detection, conspicuous signals like alarm calls warn all group members about an impending attack. Such signals allow individuals to benefit from all the eyes and ears available

Animal Vigilance. http://dx.doi.org/10.1016/B978-0-12-801983-2.00001-2

in the group to detect threats, making it possible for group members to reduce their own vigilance at no increased risk to themselves. These concepts will be explored more fully in the following chapters.

The original account of the Damara cattle story was published earlier in a relatively obscure magazine article (Galton, 1871). However, the above excerpt was part of a widely cited book on human faculty and development. It is quite surprising that early students of animal vigilance apparently ignored this work. I could only find one citation to Galton's story buried in a book on animal aggregation (Allee, 1931). It was not until the early 1970s that the work was mentioned again (Hamilton, 1971), a time when the study of animal vigilance picked up in earnest. It is perhaps the case that the message fell on deaf ears.

Ever since the 1970s, vigilance has been recognized as a major component of anti-predator behaviour. In the sequence of events leading to the eventual capture of a prey animal by a predator, vigilance plays a role in the early stages (Endler, 1991). In yet earlier stages, prey can reduce encounters with a predator by being more difficult to locate with adaptations like crypsis or aggregation (Krause and Ruxton, 2002). In later stages, prey animals can reduce the probability of capture following the launch of an attack by adopting defensive formations or confusing the predator. Vigilance plays a role in between these stages by reducing the probability that the predator remains undetected until it is too late to escape successfully. Aimed at predation threats, vigilance can be viewed as a pre-emptive measure to reduce the risk of attack because a detected predator is less likely to pursue the attack (Caro, 2005). Vigilance can also be aimed at conspecifics; in this case, it also serves a pre-emptive role by preventing or avoiding encounters with threatening individuals.

In this chapter, I lay the foundation for the scientific study of animal vigilance. First, I will provide a definition of vigilance and then pinpoint landmark studies, stretching from the pioneering work of Galton to the more recent theoretical and empirical work. I then explore research themes associated with the modern study of animal vigilance.

1.2 DEFINITION AND MEASUREMENTS

1.2.1 How to Define Vigilance

In the Oxford dictionary, vigilance is defined as the action or state of keeping careful watch for possible danger or difficulties. The Damara cattle in the Galton story, using their eyes, ears, and noses to detect ambushing lions, are clearly vigilant according to this definition. To add a biological twist to the definition, one can replace 'careful watch for possible danger' by 'monitoring the surroundings for potential threats', whose nature will be explored later in this section. While the term 'careful watch' conjures the idea that vigilance is carried out visually, the term 'monitoring' implies that all senses can be used for detection, as the cattle example illustrated.

A key feature of the definition is that vigilance can be viewed as a state or behaviour. The state of vigilance, being a predisposition of the brain, cannot be observed directly, but the outward signs, in terms of behaviour, can be observed and measured. I refer to these outward signs of a vigilant state as markers of vigilance. The dichotomy in the definition is reflected by the terms variously used over the years to describe vigilance (Table 1.1). Labels such as watchfulness, wariness, attentiveness and apprehension certainly refer to an internal state that governs how an animal monitors the surroundings for danger. Other labels describe the ways animals actually monitor their surroundings, and fall in the marker family. Terms such as head-turning, scanning, and sniffing convey the observable ways animals use their senses to detect threats. Early researchers described vigilance using terms that are now considered anthropomorphic, such as guarding and sentry-duty, which give the impression that individuals have been assigned a duty by a third party for the benefit of the group. Such terms are avoided nowadays.

In the first review on animal vigilance, vigilance was defined operationally as the probability that an animal will detect a given stimulus at a given time (Dimond and Lazarus, 1974). This definition was clearly influenced by operations research, whose goal is to study how human subjects detect important stimuli in their environment (Davies and Parasuraman, 1982). While this more psychological approach to vigilance can allow us to understand the mechanisms underlying threat detection, the emphasis on the outcome of vigilance, namely, the detection, rather than the means to achieve detection, which we can measure, makes it difficult to apply in the field.

Using the operational definition of vigilance would, however, solve the problem of actually finding a marker of vigilance because only the outcome matters. Nevertheless, measuring the probability of detecting threatening stimuli in the field remains challenging. Because predator attacks tend to be relatively rare, observational studies face the difficulty of acquiring sufficient data. Experimental production of threatening stimuli, while feasible (Godin et al., 1988; Kaby and Lind, 2003; Lazarus, 1979a; Lima, 1995a), faces the challenge of simulating convincing attacks.

But perhaps the greatest weakness in the operational definition of vigilance is that it conflates several mechanisms that can influence detection. Detection, the outcome of vigilance, is potentially influenced by three different classes of factors. The first class represents the conspicuousness of the threatening stimuli. Less conspicuous stimuli will be detected less easily, everything else being equal. For example, low light levels or visual obstacles, such as vegetation, render the detection of a predator lurking in the bushes more difficult. The second class of factors relates to the monitoring effort, the time and energy allocated to searching for dangerous stimuli. Again, all else being equal, threatening stimuli are more likely to be detected if monitoring is more intensive. Sniffing longer or visually scanning a wider area will likely increase the chances of locating a predator nearby. Finally, the brain itself, which processes all signals sent by the

TABLE 1.1 A Lexicon of Vigilance Terms

Term	Meaning	References
Anthropomorphic terms		
Sentinel	High level of alertness maintained by one individual allowing the others to forage less warily	Williams (1903)
Sentry-duty	Performing sentinel behaviour	Darling (1937)
Guarding	Maintaining a state of alertness to warn others	Jenkins (1944)
Surveillance	State of alertness for threats	Russell (1932)
Internal state terms		
Watchfulness	State of alertness for threats	Darling (1937); Russell (1932)
Attentiveness	State of alertness for threats	Tinbergen (1939)
Wariness	State of alertness for threats	Salyer and Lagler (1940)
Alertness	State of an individual on the alert for potential threats	Altmann (1951)
Vigilance	State of alertness for threats by predators	Cameron (1908); Melzack et al. (1959)
Social vigilance or social attention	State of alertness for threats by conspecifics	Chance (1967); Hall (1960); King (1955)
Fearfulness	State of fear accompanied by frequent head-turning	Tolman (1965)
Apprehensiveness	Internal state that underlies predation threat monitoring	Brown et al. (1999b)
Behavioural pattern terms		
Monitoring	Eyes, ears or noses monitor the surroundings for threats	Galton (1871)
Looking-up or around	Visual monitoring of threats	Jenkins (1944)
Scanning	Visual monitoring of threats	Leopold (1951); Lord (1956); Marler (1956)
Head-turning	Movements of the head during monitoring	Marler (1956)
Glance	Movements of the head to monitor other group members	Chance (1956)

Stimuli conspicuousness   Monitoring effort   Brain responsiveness

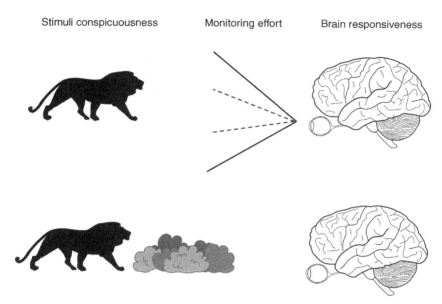

FIGURE 1.1   A framework for threat detection. Stimuli that signal danger elicit a response in the vigilant animal through the interplay of three different classes of factors: (1) factors that affect stimuli conspicuousness to the monitoring senses, (2) factors that affect monitoring effort and (3) factors that affect how the brain responds to sensory information. In the example of a lurking lion, the probability of detection by the prey animal would increase if the predator is more conspicuous (e.g. not hidden by vegetation), if more time and energy is allocated to vigilance (e.g. a greater area is scanned each time), and if the brain is more responsive (e.g. not fatigued or multi-tasking).

monitoring senses, may vary in its level of responsiveness to these signals. We have all experienced fatigue after a long car drive, which results in a decrement in vigilance performance for the same monitoring effort. The brain may also tune out when multi-tasking so that the very same stimuli that elicited a motor response earlier, say, like fleeing, are now less likely to result in a change in behaviour (Fig. 1.1). The probability of detecting a stimulus at a given time will reflect the unknown contribution of all these classes of factors. Perhaps in response to all these challenges, most studies have focused on identifying markers of vigilance rather than measuring the probability of detecting threats.

The above discussion highlights a conundrum in the study of vigilance. The state of vigilance remains unobserved unless we probe the brain to detect the state directly, which is not possible in the field. Markers of vigilance, by contrast, are measurable as distinctive behavioural patterns, such as head-cocking or sniffing (Fig. 1.2), but can be problematic as well. A good marker should consistently be observed when the animal is in a vigilant state and be conspicuously absent when the animal is not. The ideal marker should thus be both sensitive and specific to increase our confidence that when we see the outward sign of vigilance it actually means that the animal is in a vigilant state.

FIGURE 1.2 Markers of vigilance: three familiar examples of postures used to define vigilance. In meerkats, an intent gaze is used to detect threats from above. In Thomson's gazelle, one individual is feeding head down while the other is visually scanning the surroundings for threats. Notice also that the ears are pointing in one direction, possibly a sign of auditory vigilance. The head-up position is often considered a marker of vigilance. In a family of Canada geese, the male stretches its neck to monitor the surroundings, a posture thought to reflect high-level vigilance. *(Photo credits: meerkat (William Warby), geese (Andy Roberts) and Thomson's gazelles (Adrian Valenzuela)).*

One problem with many markers of vigilance is their low predictive value. There are many reasons an animal may raise its head, say, only one of which has to do with threat detection. Head-raising, for instance, can be used to locate obstacles before initiating a move or to detect prey items (Harley and Wagenaar, 2014). Animals might also appear outwardly vigilant, but may be tuned out and largely irresponsive to outside threats. As an example, many bird species sleep with one eye open and appear to keep watch at the same time. However, the time needed to detect an approaching threat is much increased during sleep, suggesting that having an eye open, the marker of vigilance, is not necessarily indicative of an internal state of vigilance (Ball et al., 1984). The reverse can also be true. In many species, individuals in a vigilant state may not show overt signs of vigilance. Rats and many bird species can detect threats even when foraging in a head-down position (Fernández-Juricic et al., 2008; Wallace et al., 2013).

To sum up, an outward sign of vigilance, such as a change in posture, might not necessarily be indicative of an internal state of vigilance, and an internal state of vigilance might not always be obviously observable. It is clear that many assumptions are made about what the senses can perceive in any measurement of vigilance based on behaviour. Nevertheless, little is known about the detection ability of many animals (Fernández-Juricic, 2012). In the end, the best markers will be those that are shown to be consistently associated with threat detection. As a final comment, I note that using markers to define vigilance has the unfortunate consequence that most studies of vigilance are restricted to birds and mammals, the two taxonomic groups where measurable changes in postures or other behavioural patterns appear related to monitoring.

1.2.2   Types of Vigilance

1.2.2.1   Routine or Induced Vigilance

The marker approach to defining vigilance implies that there is a certain level of incompatibility between vigilance and other activities because they should be sufficiently different in outlook to be clearly separated into different activities. Few species are as fortunate as the pram bug when it comes to maintaining vigilance. The pram bug is a small but scary looking amphipod of the deep seas that may have been an inspiration for the monster in the Alien movies. This species possesses two pairs of compound eyes each set of which faces in a different direction thus providing the animal with a 270° view of its surroundings. This wide field of view allows individuals to simultaneously detect prey in front of them and any approaching predator from nearly all sides (Ball, 1977). More typically, prey animals must trade-off vigilance and the time, energy and attention devoted to other activities, such as foraging and sleeping. Individual allocation of time and effort to vigilance is widely viewed as a balancing act between competing activities (Caraco, 1979b; Lima, 1987b).

In the trade-off view of vigilance, two types can be distinguished: routine or induced vigilance (Blanchard and Fritz, 2007). Routine vigilance takes place when an individual monitors its surroundings during spare time. Induced vigilance is more costly as it disrupts the foraging process or other activities. Here, vigilance is characterized by the willingness to incur costs, with the expectation that costlier actions reflect a greater investment in vigilance and thus a greater probability of detecting threats (Brown et al., 1999b). For instance, a more vigilant gazelle that completely stops foraging to investigate a potential threat will experience a larger decrease in food intake rate than a less vigilant one that simply scans the surroundings while processing food. Other terms to characterize vigilance according to the level of investment include low- and high-level vigilance (Lazarus, 1979a) or subtle and overt vigilance (Monclús and Rödel, 2008).

Classifying vigilance is not always immediately obvious. In the blue tit, a small European forest passerine bird, individuals detected simulated attacks to the same extent whether engaging in so-called low- or high-level vigilance (Kaby and Lind, 2003), suggesting that the two types of vigilance are one and the same as far as threat detection is concerned. In other species, the extent to which individuals stretch their necks during visual monitoring was used to distinguish between low- and high-level vigilance (Lazarus, 1978). The classification of vigilance levels based on associated costs is less obvious if vigilance does not interfere with other activities like foraging. Some herbivores can overlap foraging and vigilance by chewing their food head up. Here, induced vigilance might be inferred when animals interrupt food handling to investigate a potential threat (Baker et al., 2011; Cowlishaw et al., 2004; Favreau et al., 2013; Fortin et al., 2004a).

When possible at the phenotype level, a distinction between vigilance types based on costs can be useful because the different types of vigilance might be adjusted independently. The costlier form of vigilance might be used preferentially in contexts deemed more threatening. In impalas, an ungulate of the African savannah, males and females, which face different types of threats, showed a different mixture of the two types of vigilance when exposed to threats more significant to one sex or the other (Favreau et al., 2013), highlighting the usefulness of distinguishing the two types of vigilance.

1.2.2.2 Anti-Predator or Social Vigilance

Vigilance can also be broken down according to the intended target, resulting in two types of vigilance: anti-predator and social. Obviously, the senses of an animal can be directed at other targets, such as obstacles on the path ahead or prey items. As mentioned earlier, alternative targets make markers of vigilance less specific. In this book, I am only concerned with potentially threatening stimuli, and leave aside scanning for prey items, inanimate objects or any environmental or social stimuli that do not pose a threat.

Anti-predator vigilance is aimed at detecting the presence of a nearby predator by using subtle signs like rustles in the vegetation or a silhouette in the sky. Social vigilance, a term first coined in the early 1960s (Table 1.1), recognizes that danger can also stem from nearby conspecifics. Chacma baboons in the African savannah, the species for which the term was first coined, divide their vigilance between predators and conspecific foes (Hall, 1960). Social vigilance is directed at the more dominant group members who use aggression to displace others from preferred locations or food patches (Chance, 1956). Other targets of social vigilance include mating rivals (Li et al., 2012; Rieucau et al., 2012), individuals that steal food items (Goss-Custard et al., 1999) or those that can kill offspring (Steenbeek et al., 1999). The distinction between these two types of vigilance is important because each can respond differently to the same ecological pressures. In the next chapter, we will see, in particular, that anti-predator vigilance is expected to decrease with group size while social vigilance typically increases.

How to practically distinguish between anti-predator and social vigilance is not always simple. In species with eyes facing forward, like in some primates and ungulates, the direction of the gaze can provide a clue as to the intended target of vigilance. In a study of Eastern grey kangaroos, researchers inferred social vigilance when individuals at the periphery of the group gazed towards other group members and anti-predator vigilance when they looked away from the group (Favreau et al., 2010). In animals with more laterally placed eyes and broader fields of view, it may not be so easy to determine the target of vigilance. Consider that when the bill points forward in chickens, a species with laterally placed eyes, the birds may actually be looking sideways (Dawkins, 2002). Recent eye-tracking technology may be quite useful to determine what animals are actually looking at and to develop useful markers of gaze direction (Yorzinski and Platt, 2014).

The two types of vigilance can also be distinguished statistically by pitting putative correlates of predation and social risk against one another in the same model to determine which set of variables is most closely associated with empirical levels of vigilance. This approach allowed researchers to conclude that most of the vigilance in spotted hyenas was directed at conspecifics rather than predators (Pangle and Holekamp, 2010b).

1.2.2.3 Pre-Emptive or Reactive Vigilance

The typical view of vigilance is that it allows individuals to detect threats of any sorts early. Vigilance is thus a pre-emptive strike providing extra time to make proper adjustments in response to a threat. Reactive vigilance, by contrast, is concerned with the actual monitoring of a detected threat (Boinski et al., 2003; Hoffman et al., 2006), be it a predator or a competitor. Reactive vigilance takes place, for instance, during encounters between rival groups in territorial primates (Gaynor and Cords, 2012; Macintosh and Sicotte, 2009) or when a predator has just been detected (Li et al., 2009; Wichman et al., 2009; Yaber and Herrera, 1994).

While monitoring the behaviour of a nearby predator still represents vigilance, as broadly defined earlier, other considerations come into play during predator monitoring. The duration of vigilance at this stage has been related to factors such as the current state of the prey animal and the distance between the predator and the prey (Ydenberg and Dill, 1986). As such, vigilance directed at the detected predator can be viewed as the next stage in the predation sequence, one that is concerned with the timing of fleeing (Caro, 2005). In this book, I will be mostly concerned with vigilance before the threat has been detected.

1.2.2.4 Visual, Auditory, Olfactory or Vibrotactile Vigilance

The last classification scheme emphasizes the senses involved during vigilance. Most of the examples described thus far focused on visual vigilance. Auditory vigilance, however, is probably widespread in animals but harder to document. Herbivores, for instance, are thought to stop handling food to reduce food-processing noises that can mask auditory cues associated with approaching danger (Blanchard and Fritz, 2007). Observations that animals become visually more vigilant in noisier settings certainly suggest that auditory vigilance plays a role in detecting danger (Quinn et al., 2006; Rabin et al., 2006). The difficulty with auditory vigilance is to determine when it occurs. Markers of visual vigilance have been discussed previously, but comparable markers of auditory vigilance are not well-defined. In species without external movable ears, only head movements may betray auditory vigilance, but then again such movements may be associated with visual vigilance. Changes in the orientation of the ears and other movements such as ear twitching might indicate auditory vigilance (Basile et al., 2009), but I am not aware of any studies that have related observable ear movements to predation risk. Since visual and auditory vigilance may be carried out simultaneously, an experimental approach is probably needed to disentangle their relative contribution to overall vigilance.

Olfactory vigilance, which is concerned with the detection of smells associated with predators or predation events, is well established in fish and mammals (Apfelbach et al., 2005; Ferrari et al., 2010). In fish, chemical substances released by mechanical breakage of the skin after a predation event are known to alert other fish about the recent presence of a predator (Krause, 1993). Olfactory vigilance of this type may thus be useful to avoid risky habitats or to signal risky times. In mammals, sniffing for predator smells has been documented in many species (Blumstein et al., 2008; Kuijper et al., 2014), including the cattle in the Galton story. There is also evidence that birds may be able to use smells to assess predation risk from mammals (Roth et al., 2008a). Whether olfactory vigilance carries costs like visually induced vigilance is not clear, but if sampling smells interferes with foraging, say, it might be possible to predict the optimal allocation of time to olfactory vigilance.

Another type of vigilance has been related to vibrations in the substrate caused by the movements of predators. Effective detection of vibrations caused by predators, in contrast to those produced by abiotic factors or conspecifics, has been documented in one caterpillar species (Castellanos and Barbosa, 2006). While it may not be possible to ascertain from its demeanour when a caterpillar is vigilant for such vibrations, the outcome of the detection process can be easily documented from the subsequent changes in posture. Vibrotactile vigilance is mostly known in invertebrate species (Hill, 2009), but since many other types of species including mammals use vibration-based communication channels (Randall, 2001), detection of predators through vibrations in the substrate may be more widespread than expected. In one of the most famous scenes in the movie Jurassic Park, the lumbering approach of *Tyrannosaurus rex* was indeed betrayed by ripples visible on the water.

1.2.3 How to Measure Vigilance

1.2.3.1 Measuring the State of Vigilance

Indirect indices have been sought to infer a vigilant state when external signs of vigilance are not obvious. Detection of a threat, which is indicative of a state of vigilance, has been related to skittering in fish (Magurran et al., 1985) and to changes in skin colour in squids (Mather, 2010), but this does not say anything about the occurrence of a vigilant state when a threat is absent. Another approach is to use a specific marker for a behavioural pattern unrelated to vigilance and to assume that vigilance occurs when that marker is absent. For instance, vigilance was scored when an animal was not sleeping (Lanham and Bull, 2004) or not fighting (Brick, 1998), two states that are easily characterized but whose opposites may not be very specific to vigilance.

For species with no specific markers of vigilance, the only possibility left to infer a vigilant state is to determine the probability of detecting a threat. As I discussed previously, an inference about the level of vigilance from the probability of detection makes implicit assumptions about stimuli conspicuousness, monitoring effort, and brain responsiveness. This approach is mostly experimental and consists in producing a stimulus thought to be representative of predation risk and then noting how much time is needed to show an overt reaction (Fig. 1.3). Examples of stimuli used in various experiments include a brief flash of light (Godin et al., 1988), a brief sound (Lazarus, 1979a), artificial vibrations produced mechanically (Castellanos and Barbosa, 2006), a small ball rolling towards a targeted bird (Lima, 1995a), and stuffed predators on a wire (Kaby and Lind, 2003). The time needed to respond to these stimuli is thought to indicate the level of vigilance of the animal with longer delays betraying a lower investment in vigilance.

One caveat of this type of approach is that the stimuli used to simulate predation are not always particularly realistic. This may have been the case in a recent study that assessed adjustments in migratory stopover behaviour in small shorebirds in response to variation in predation risk. In this experiment, a crude silhouette of a falcon was shot from a bow, but often failed to elicit any reaction from the flocks (Hope et al., 2014). A second limitation is that an overt reaction may be delayed adaptively, reflecting a trade-off between foraging gains

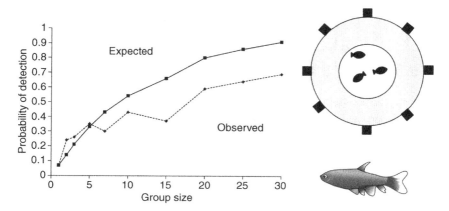

FIGURE 1.3   Detection in groups: the probability of detecting a frightening stimulus (a brief flash of light) increased with shoal size in the glowing tetra fish. The empirical probability of detection (dashed line) was smaller than expected under the assumption of independent stimulus detection (solid line), suggesting that vigilance became more relaxed in larger groups. Fish were held in an aquarium and the brief flash of light could originate from any of eight positions at the periphery of the tank. *(Adapted from Godin et al. (1988)).*

and fleeing from a threat (Ydenberg and Dill, 1986), which makes it difficult to determine the exact time of reaction.

1.2.3.2 Measuring the Markers of Vigilance

The markers of vigilance are typically concerned with visual vigilance. For such markers, estimates of vigilance are usually gathered from time budgets. In a time-budget study, the percentage of time allocated to vigilance during a predetermined time period is obtained from a number of different animals. In groups of animals, another possibility is to estimate the percentage of individuals in the group that are vigilant at predetermined time fixes and to aggregate the percentages over the duration of the observation. This technique is suitable to get an idea of the time allocated to vigilance by an average group member and produces one estimate of vigilance per group.

With focal sampling, by contrast, three measures of vigilance can be calculated after pooling results from all focal individuals: time spent vigilant, which is the average percentage of time allocated to vigilance during a bout of observation, vigilance frequency, which is the average number of vigilance bouts initiated per unit time, and vigilance duration, which is the average duration of all vigilance bouts (Fig. 1.4). These different measures tend to be correlated with one another, but they are not interchangeable and can be adjusted independently (Beauchamp, 2008b).

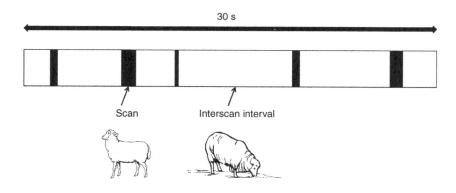

Frequency of scanning per min: $5 \times 60/30 = 10$

Mean duration of scanning: $(0.75 + 1 + 0.25 + 0.75 + 1)/5 = 0.75$ s

Time spent scanning: $(0.75 + 1 + 0.25 + 0.75 + 1) \times 100/30 = 12.5\%$

FIGURE 1.4 How to measure vigilance: raising the head during browsing represents a potential marker of vigilance in sheep. Scanning the surroundings for threats takes place when the head is up. Inter-scan intervals include periods when the head is down. Three different estimates of vigilance can be obtained from a focal observation, here lasting 30 s: the frequency of scanning (number per minutes), the mean duration of a scan (in seconds), and the percentage of time spent scanning.

Solely focusing on time spent vigilant can hide crucial information about vigilance. The same percentage of time spent vigilant can be achieved by a few long scans or many short ones (McVean and Haddlesey, 1980). The different ways to achieve the same level of vigilance overall are not necessarily equivalent. For instance, infrequent but relatively long scanning bouts are probably more useful than frequent short ones when predators move slowly and the direction of attack is quite unpredictable. Which particular combinations of vigilance duration and frequency should be preferred has received little attention (Sirot and Pays, 2011).

Early studies often viewed vigilance as a simple pattern of raising and lowering the head. Subtle head movements and even eye movements while actually scanning the surroundings can be used to optimize detection or monitor different targets, such as predators or companions (Fernández-Juricic et al., 2011a; Jones et al., 2007; Wallace et al., 2013). These more subtle adjustments suggest that solely focusing on the duration of a monitoring bout may be a poor proxy of the quality of the information acquired.

1.3   A HISTORY OF VIGILANCE

Here, I provide a brief history of vigilance research. This historical perspective highlights major developments over the last 150 years or so, at least as I see it (Fig. 1.5). It is not meant to be an exhaustive list of all relevant references, but rather an informal look at the past to understand current trends in vigilance research.

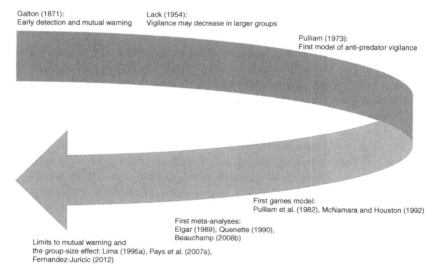

FIGURE 1.5 **A history of vigilance.** Key stages in the history of research on vigilance from Galton (1871), who first articulated the concept of mutual warning in a group and emphasized the role of vigilance in the early detection of predators, to recent research examining how causal factors modulate vigilance. Other key stages include theoretical developments, which allowed the formulation of qualitative and quantitative predictions, and meta-analyses, which reviewed particular predictions in birds and mammals.

1.3.1 First Phase

The first descriptions of vigilance behaviour, and its potential role in predator detection, go back to roving British naturalists of the late nineteenth century. Henry Bates, for one, visited the Amazonian jungle in the 1860s and achieved fame for describing how harmless species of butterflies evolved to mimic the warning signals of less palatable species, thereby gaining protection against their common predators. He also noticed the occurrence of large flocks of birds composed of many different species and speculated that individuals in a flock with so many eyes would be less likely to be surprised by a predator (Bates, 1863), yet another mechanism to increase safety in prey animals. Belt reached the same conclusion a few years later based on his own observations of mixed-species flocks of birds in Nicaragua (Belt, 1874).

Galton (1871) espoused a similar view to explain gregariousness in the Damara cattle, as described in the introduction to this chapter. Galton both described and speculated on the function of vigilance in cattle. The position of the head, down when grazing and up when monitoring, represents a marker of vigilance in this species. Galton assumed that an ox grazing amidst vegetation cannot easily detect a threat. He surmised that all senses can be used to monitor threats, including the eyes, ears and nose. The function of this vigilance is to avoid being surprised by a lion in ambush. The ox in a group also enjoys an advantage over a lone beast because of the multiplicity of senses tuned to predator detection, making it less likely that a threat will go unnoticed. This remarkable description was either ignored or not cited by subsequent researchers interested in vigilance behaviour, and only resurfaced about 100 years later (Hamilton, 1971).

The first phase of vigilance research stretches from these early observations to the early 1970s. During this phase, researchers typically provided accounts of vigilance in their study species and speculated on its function. I found the first mention of the word vigilance, as understood in animal behaviour, in a paper describing how golden eagles high in the sky launched attacks on prairie dogs below when individuals relaxed their vigilance (Cameron, 1908). A distinction was made early between the state of vigilance, a reference to an internal state that governs monitoring, and the act of vigilance, the visible ways animals monitor their surroundings. Words such alertness (Altmann, 1951), watchfulness (Darling, 1937; Russell, 1932), attentiveness (Tinbergen, 1939) and attention (Chance, 1967) describe the state of a prey animal in threat detection mode. Sniffing (Galton, 1871), looking around (Jenkins, 1944) and scanning (Leopold, 1951; Lord, 1956; Marler, 1956) describe how the animals pay attention to their surroundings.

The function of vigilance was originally thought to be the early detection of predators. This alleged function reappeared in many subsequent accounts of the adaptive value of group living in animals although it is not clear whether researchers reached this conclusion independently given the lack of reference to earlier work in most papers. In an early review of the adaptive value of flocking in birds, Miller thus proposed that a flock offers the advantage of a multiplicity of eyes on the lookout for predators (Miller, 1922). Given that Galton had earlier suggested the idea that a group benefits from the multiplicity of senses, it may be the case that Miller adopted the same terminology after reading his account, but this is just speculation. I also noted a reference to a multiplicity advantage in Allee's influential book on group living in animals published 10 years later (Allee, 1932).

Some researchers argued that early detection through vigilance was not the main reason animals benefited from being in a group. The confusion experienced by a predator faced with the fleeing of many individuals at the same time was in fact thought to be more effective than early detection (Winterbottom, 1949). Others actually argued that protection from predators through vigilance or any other mechanisms played only a small role in the evolution of sociality (Murton, 1971; Rand, 1954). Hard evidence for or against these hypotheses was not provided.

Another recurrent theme in vigilance research originated from this early phase. In the Galton cattle story, the start or a cry from one individual after detection soon alerted the whole group. The ability to react to alarm calls or fright reactions produced after detection of a threat is known as mutual warning or collective detection. Allee (1932) also makes a reference to this mechanism. Without mutual warning, group members cannot benefit from the multiplicity of senses and must rely on their own detection ability. This important mechanism is explored more fully in Chapter 5. The ability to warn less wary group members also forms the basis of sentinel behaviour in animals. In a sentinel system, one or more individuals maintains a high level of vigilance, typically from a vantage point, and their alarm calls alert less vigilant companions (Elliott, 1913; Williams, 1903). Sentinel behaviour is of course well-known in our own species. It recurs in countless action movies in which weary-eyed cowboys or pirates take turns keeping watch for marauders while the others sleep.

In his influential book on adaptation and natural selection, Williams sketched the idea that individuals in a group could benefit by using others as a shield against predators (Williams, 1966). This idea was to become the foundation of the selfish-herd mechanism developed a few years later (Hamilton, 1971). He also speculated on other adaptations that would benefit animals in groups. Apparently unaware of the earlier descriptions of mutual warning, Williams suggested that the ability to recognize signs of alarm in other group members and to act defensively in response would be of great adaptive value.

A final theme also emerged during the first phase of vigilance research. Galton, in the cattle story, suggested that in a group an individual can receive a maximum of security at a minimum cost of restlessness. If restlessness represents the time taken away from foraging by anti-predator vigilance, then he was indeed suggesting that an individual in a group can achieve a high level of safety despite not being overly vigilant. The idea that the level of vigilance maintained by individuals may be reduced in a group at no increased risk to themselves forms the basis of the group-size effect on vigilance, which will be a defining theme of the second phase of vigilance research (Section 1.3.2).

A less obtuse, perhaps independent, formulation of the group-size effect on vigilance was provided by the eminent avian biologist David Lack in the mid-1950s. In his book on the regulation of animal numbers, Lack surmised that that an animal feeding alone would have to spend more time alert and less time feeding to achieve the same level of safety as individuals in a group (Lack, 1954). This advantage stems from the multiplicity of senses available in a group, which should allow individuals to relax their vigilance and still reap the benefits of extra safety. In his studies of wood pigeons feeding in agricultural fields in England, Murton and his colleagues provided the first evidence for this phenomenon when they noted that solitary birds spent more time looking around and fed less than their counterparts in groups (Murton, 1968). The first demonstration that vigilance systematically decreases across a large range of group sizes was provided by Lazarus. This work was mentioned in passing in a letter published in Ibis (Lazarus, 1972), discussed some more in the first review of animal vigilance (Dimond and Lazarus, 1974), and eventually published later (Lazarus, 1978).

1.3.2 Second Phase

While the first phase of vigilance research emphasized the description of vigilance behaviour and suggested ways animals benefit from spending time alert, there was no clear research agenda until the early 1970s. Reflecting the lack of organized research on vigilance, Welty, in his widely read book on the life of birds, provided no mention of vigilance at all in the 1962 edition of the book. In the second edition, in 1968, he suggested in a very brief section that flocking in birds can reduce predation risk by allowing early detection of predators through vigilance (Welty, 1968). It was quite clear that there was little interest in the systematic study of vigilance at that time.

The picture changed drastically in the early 1970s. Reviewing the literature on the adaptive value of flocking in birds, Murton, who had earlier provided some of the finest observations on vigilance and early detection of predators in birds, flipped his position and argued that feeding in groups probably had no anti-predator advantages (Murton, 1971). In a rebuttal published subsequently, Lazarus, then a young PhD student, suggested that Murton downplayed the importance of predation and pointed out that individuals in groups may be able to allocate more time to feeding because of the increased safety afforded by the multiplicity of alert senses (Lazarus, 1972).

Not 1 year later, in a short note published in a theoretical journal, Pulliam provided the first mathematical model of vigilance in prey animals (Pulliam, 1973) apparently unaware of the work of Lazarus or Lack on the same topic. In any case, the work was the first to formalize ideas about vigilance. In particular, the model provided a function to calculate the probability that a predator will be detected by a prey animal that alternates bouts of vigilance, during which predator detection takes place, and bouts of feeding, during which detection is impossible. Pulliam also showed that a reduction in individual vigilance is possible in a group without reducing safety. The two key predictions from this model, namely, that detection of threats occurs sooner in larger groups and that individual vigilance can decrease in larger groups, set the agenda for much of the later empirical research.

It is no coincidence that research on vigilance started in earnest during this period. There was considerable interest at the time on the anti-predator function of animal aggregations (Crook, 1965; Goss-Custard, 1970), which culminated in the publication of a book on social behaviour in birds and mammals (Crook, 1970). Influential models showing how aggregation can reduce predation risk also stimulated discussion on other potential advantages of group living in animals including the detection of predators through vigilance (Hamilton, 1971; Vine, 1971, 1973).

Vigilance, as a topic of research, certainly attained a new status by the mid-1970s. In their widely read syntheses of research on animal behaviour, both Alcock and Wilson mentioned the use of vigilance to detect predators early (Alcock, 1975; Wilson, 1975). Furthermore, Wilson stressed the fact that solitary animals need to spend more time vigilant. In the first edition of their widely successful edited book on behavioural ecology, Krebs and Davies included one chapter on living in groups that highlighted the role that vigilance plays in avoiding predation (Bertram, 1978).

The second phase of research on vigilance saw many new theoretical developments, which will be described in more details in Chapter 5. While these developments were essential to provide a more satisfactory treatment of vigilance in groups (McNamara and Houston, 1992; Pulliam et al., 1982; Sirot, 2012), few of their novel predictions have been tested. The lion's share of research has been devoted to the group-size effect on vigilance, with literally hundreds of paper published in the four decades that followed the pioneering work of Pulliam. Large-scale reviews in birds and in mammals (Beauchamp, 2008b, 2013b; Elgar, 1989; Quenette, 1990) have synthesized the available information. This strand of research will be reviewed in Chapter 6.

Research on vigilance in the second phase has focused primarily on functional explanations of the behaviour. Explanations that emphasize causal mechanisms have emerged more recently (Fernández-Juricic, 2012). Causal research asks how vigilance is achieved and how the senses constrain the information gathered during vigilance. There has only been marginal interest in developmental questions, or how vigilance changes during the lifetime of an individual, or in the evolution of vigilance (Beauchamp, 2014b; Caro, 2005). These questions are tackled in Chapter 3.

1.4   RESEARCH THEMES

How would we describe the field of vigilance research to a student ready to embark on a vigilance project? Definition issues, reviewed in Section 1.2, can be of practical help in devising a research protocol. The historical perspective, presented in Section 1.3, can be useful to see how vigilance research evolved over the years. But this says little about current research trends or whether vigilance research remains relevant.

Over the years, I have duly collected references from the literature dealing with various aspects of vigilance (close to 1300 as of December 2014). There has been an exponential-like increase in the number of publications dealing with vigilance over the years (Fig. 1.6). However, before claiming that vigilance has become more popular with animal behaviour students, it is important to realize that the total number of publications dealing with animal behaviour in general (to be precise, the number of publications reported by Google Scholar including any of the following terms: animal behaviour, behavioural ecology, behavioural ecology and sociobiology or ethology) has increased just as dramatically during the same period. This increase partly reflects the greater need

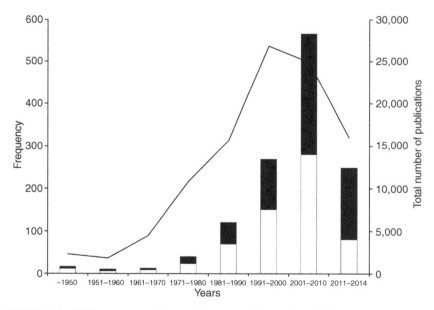

FIGURE 1.6   Vigilance research through the years. The number of published papers on vigilance has risen sharply through the years, especially after the publication of Pulliam's paper (1973), which predicted that individuals in groups could reduce their own vigilance at no increased risk to themselves. Early research typically focused on bird species (proportion shown in white in each bar), but research on mammals and other species (proportion shown in black in each bar) has become more prominent recently. Research output by students of animal vigilance kept pace with the total research output in the fields of animal behaviour and behavioural ecology (line graph).

to publish in science in general, the larger pool of available researchers, and also the increase in the number of journals where vigilance research can be published. With respect to the last point, research on vigilance before 1970 was typically published in *Animal Behaviour*, the sole journal that specialized on animal behaviour at the time. Nowadays, research on vigilance can find a niche in a dozen journals or so.

In this inflationary context, what is perhaps more telling is the proportional share of effort allocated over time to vigilance research. This proportional share increased over the years, especially after the early 1970s where a clearer research agenda emerged following Pulliam's publication (Fig. 1.6), indicating that research on vigilance remains active. While early research on animal vigilance typically focused on birds, the effort allocated to non-avian species (mammals and others) has increased over the years, indicating a healthy diversification of the study species (Fig. 1.6).

The field of vigilance research continues to thrive, but this tells us little about the current research themes embraced by animal behaviour students. I certainly have formed an impression of the more obvious themes by reading the literature on this topic over the last 20 years, but I am probably biased because some themes are more appealing to me than others. To get a more objective view, I used standard text data mining to determine which terms come up frequently in the titles and abstracts of the large pool of references that I used to produce the previous figure. Text data mining leaves aside common terms, such as prepositions, and allows one to get a picture of the frequency with which terms appear together in a reference. Such co-occurrences can help us get a sense of what animal vigilance students investigate. Obviously, there are limits to this text mining approach. Abstracts are not always available for older papers. It may also be difficult to infer a research theme just by one word.

With these caveats in mind, the text mining exercise revealed interesting features (Fig. 1.7). Primarily, vigilance is studied in the context of predation, not surprisingly, and foraging. Research on vigilance also often involves groups. This trend is probably underestimated since the term 'group' has many synonyms, such as flock or herd, in different types of animals. A research project on vigilance thus typically investigates anti-predator vigilance in foraging groups.

What about the types of animals used in vigilance research? The most popular organisms in vigilance research are birds and mammals, with the majority of papers mentioning birds (Fig. 1.8). This is not surprising given that Pulliam's model focused on avian species, which made them obvious targets for subsequent vigilance studies. Interestingly, the share of papers on vigilance in mammals has increased over the years (Fig. 1.6), suggesting that mammalian researchers are catching up. Families of animals are often mentioned in the papers on vigilance. In mammals, the most common types of species are monkeys and squirrels, with particular species such as elk, sheep, and marmots also taking a large share. In birds, the most common families include shorebirds and ducks, and the most common species are tits and sparrows. Surprisingly, fish

Predation

Prey

Predator

Foraging–feeding

Risk

Food

Vigilance

Group

Flock

Size

FIGURE 1.7   Themes in vigilance research. Text data mining of a large number of bibliographic references dealing with vigilance revealed important themes. Font size is proportional to the number of times a term was found in the title and abstracts of these references.

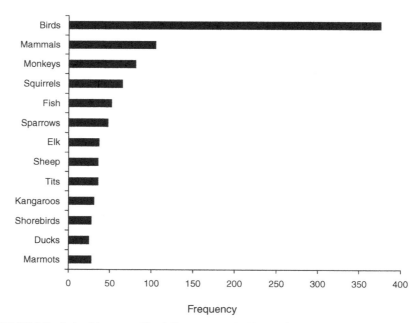

FIGURE 1.8   Animal types used in vigilance research. The text data mining exercise based on bibliographic references on vigilance showed how frequently particular types of animals appeared in these references.

came up quite frequently in the vigilance literature. As mentioned earlier, few studies in fish focus on markers of vigilance. Therefore, vigilance projects in fish probably concentrate on threat detection. In a nutshell, a typical vigilance study focuses on anti-predator vigilance in foraging groups of species like sparrows or monkeys.

1.5 CONCLUSIONS

The study of vigilance has made remarkable strides over the years and came of age in the early 1970s following the publication of Pulliam's model. Adaptive explanations of behaviour, which blossomed during this period as witnessed by the development of optimal foraging theory (Stephens and Krebs, 1986), found a perfect outlet in vigilance research. Predation avoidance through vigilance was indeed thought to be one of the main reasons behind the evolution of group living in animals (Pulliam and Caraco, 1984). I investigate these functional issues in the following chapter. It has also become quite clear over the years that vigilance is aimed not only at predators, but also at rivals and competitors, which are a much more immediate threat than predators in many species.

How to measure vigilance accurately and what constitutes the target of vigilance really represent the soft belly of vigilance research. In many cases, I am afraid, it is not clear at all when an animal is vigilant and, when it is, what it pays attention to. As a case in point, a recent study suggested that scanning in rats helps to form a cognitive map that is used later to retrieve resources (Monaco et al., 2014). In this case, the target of vigilance is neither a predator nor a foe, but rather physical features of the environment associated with resources. This type of vigilance has been linked to physiological and measurable changes in the brain compatible with the formation of episodic memory. This does not mean that vigilance in rats is never aimed at threats, but it is certainly the case that other functions of vigilance are conceivable unless the target of vigilance can be readily identified.

The recent emphasis on mechanistic explanations of vigilance will no doubt contribute to shed light on the sensory ecology of many species used in vigilance research. Instead of a black and white definition of when an animal is vigilant, we might be able to express vigilance as a continuous variable with many shades of grey. It may also prove easier to determine what the animals really are looking at or listening to during vigilance.

The lack of specific markers of vigilance in some species need not imply that avoidance of predators or detection of foes is unimportant. Conspicuous head or body movements in some invertebrate species are probably good candidates for markers of vigilance (Harley and Wagenaar, 2014), but lack of research on anti-predator vigilance in those species hampers our understanding of the ways animals detect threats. Focusing on other senses than vision will also facilitate the application of vigilance research to other types of species, including fish and invertebrates.

Chapter 2

Function of Animal Vigilance

2.1 INTRODUCTION

The chital, also known as the axis deer, is a mid-size deer found primarily in the Indian peninsula. This species of deer lives in large, fluid groups and faces some of the fiercest known terrestrial predators. In the Gir Forest National Park, in northwestern India, chitals are hunted by the few remaining Asiatic lions (Sundararaj et al., 2012). In other habitats, tigers and leopards, no slouch predators themselves, commonly prey on chitals. Faced with such a wide range of predators, chitals evolved the ability to recognize and respond to the alarm calls of many other species, including langurs and lapwings (Schaller, 1967). Langurs, in particular, forage high in the trees and can spot predators earlier than the deer, which forage low on the ground. Chitals also produce an effective array of signals (visual, auditory, and olfactory) that can be used by other group members to rapidly identify a threatening situation.

In his book on tigers and their prey in India, George Schaller provided a vivid description of sophisticated anti-predator behaviour in the chital (Schaller, 1967). A vigilant chital stands erect and stretches its neck in a clearly visible sign of alertness. In addition, an alarmed chital can raise its tail exposing the white rump underneath, which sends a strong visual signal of alarm to the nearby deer that might have missed the approaching predator. Alarm responses spread rapidly in the group and allow all group members to initiate a fleeing response much sooner than would be the case if each deer responded independently. Recent work showed that chitals decrease their vigilance in large groups, but still enjoy extra safety as the corporate level of vigilance increases with group size (Ghosal and Venkataraman, 2013). Ever sensitive to predation risk, chitals reduce their vigilance in areas where predators are less likely to attack (Sundararaj et al., 2012).

Vigilance behaviour in the chital can be explained at four different, complementary levels, as is the case with any behavioural pattern (Tinbergen, 1963). An ultimate explanation focuses on the adaptive value of behaviour. Specifically, how can vigilance contribute to survival and reproduction? For vulnerable chitals, an ultimate or functional explanation might be that fine tuning of responses to alarm signals and high level of vigilance evolved to facilitate early detection of predators and reduce predation risk. The observation that vigilance decreases in areas with low predation risk certainly fits this adaptive scenario because vigilance probably interferes with the acquisition of food, another

Animal Vigilance. http://dx.doi.org/10.1016/B978-0-12-801983-2.00002-4

major component of fitness. Other targets of vigilance are also possible. In the harem reproductive system that characterizes this species, males compete fiercely for mating opportunities and probably monitor each other extensively. Hence, vigilance can also serve the function of assessing or monitoring rivals to increase reproductive success.

A proximate explanation identifies internal or external factors that drive the actual level of vigilance. In chitals, as in other species, vigilance at this proximate or causal level might be related to testosterone levels in males (Fusani et al., 1997) or corticosteroids in general (Mateo, 2007). Still at the proximate level, the temporal structure of vigilance behaviour in the chital, namely, how frequently individuals alternate bouts with the head up and bouts with the head down, could be related to the acuity of their vision (Fernández-Juricic, 2012) or the limits of their other senses.

The level of vigilance adopted by individuals can change during their lifetime, providing an explanation of vigilance that emphasizes the development of the behaviour. In some species, individuals need to learn to associate potential threat signals in their habitats with predators (Griffin et al., 2001), which probably explain why vigilance becomes more targeted over time. Changes in vigilance levels as a function of age can also reflect maturation of the senses or the gradual acquisition of foraging skills (Alberts, 1994; Arenz and Leger, 2000), which need to be understood to allow a proper explanation of the behaviour. Development of vigilance in the chital has not been investigated yet.

A final explanation relates the level of vigilance in a species to broad ecological factors. An evolutionary explanation uses an historical perspective to explain differences amongst species in their typical vigilance behaviour. Assuming a genetic contribution to phenotypic variation in vigilance behaviour, a combination of ecological factors probably favoured in the past the evolution of the particular vigilance pattern exhibited today by chitals. High levels of vigilance in chitals could be related to their preference for largely open habitats and their relatively small size, both of which have been related to the evolution of a suite of anti-predator defences in mammals (Caro et al., 2004).

These various explanations for vigilance behaviour are summarized in Fig. 2.1. In this chapter, I will focus on functional or ultimate explanations of vigilance behaviour, leaving aside alternative explanations for the following chapter. It is important to remember that these different types of explanations are not mutually exclusive but rather complementary. Their combination provides a much fuller account of behaviour.

2.2 TYPES OF THREATS

The function of vigilance differs according to the target of vigilance and ultimately depends on the type of threats faced by an individual. Threats can be classified in two broad categories: (1) threats to survival and (2) threats to reproductive success.

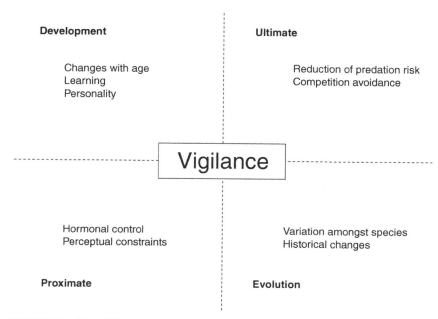

FIGURE 2.1   Four different explanations of vigilance. At the ultimate level, the focus is on the adaptive value of behaviour. For instance, does vigilance against predators increase survival? At the proximate level, the focus is on internal or external factors that modulate vigilance. For example, does the level of testosterone in males affect vigilance? Developmental questions focus on factors that alter the expression of vigilance during the lifetime of an individual, including learning and maturation. At the evolutionary level, the emphasis is on broad ecological factors that have shaped vigilance patterns in different species over evolutionary times. These four different explanations are complementary.

First, let me provide examples of each type of threats. Predators obviously pose an immediate threat to the survival of a prey animal. As we saw earlier, the life of a chital is chiefly a long struggle to avoid becoming someone else's lunch. It comes as no surprise that vigilance patterns in animals can be shaped by predation risk. However, recent work suggests that competition can also play a role. Competition is a non-lethal form of interaction at least in the short-term; it has a more indirect relationship to survival than predation (Sutherland, 1996). Individuals compete for many types of resources that can affect survival. Competition for food is a familiar example.

One predation attempt can obviously end life rather abruptly. Competition can also have nefarious consequences, but typically over a much longer time frame. Eurasian oystercatchers often steal mussels from one another when foraging (Goss-Custard et al., 1999). Losing a mussel to a competitor has no immediate effect on survival, but cumulative losses over a long period of time probably reduce fat reserves and increase the chances of dying from a cold spell or disease. Vigilance while handling large mussels would thus be useful to assess whether conspecifics are likely to attack. Monitoring predators and competitors that pose a threat to survival is therefore a major function of vigilance in animals.

I turn now to the other type of threats. Competitors can certainly pose a threat to survival, but can also jeopardize reproductive success. During rutting season, male Przewalski's gazelles in the steppes of northwestern China spend an inordinate amount of time monitoring one another in their attempts to mate with females in their groups (Li et al., 2012). Rival males compete for mating rights and vigilance is useful in this context to safeguard monopolization rights or to avoid attacks by more dominant males. Protection of paternity could also explain vigilance in other species in which females might mate with other males during their fertile phase (Artiss and Martin, 1995). Females are also involved in this type of vigilance. Vigilance in females can be used to detect or avoid harassing males or those intent on killing offspring (Steenbeek et al., 1999). Monitoring rivals and foes that pose a threat to reproductive success is the other major function of vigilance.

2.3 MONITORING COMPETITORS

Monitoring competitors is an important function of vigilance during both the reproductive and non-reproductive seasons. As the above examples amply illustrated, competitors can threaten both survival and reproductive success. Here, I explore the specific ways vigilance can help to detect such threats starting with food competition.

2.3.1 Contest Competition for Food

Competition for food is often intense, and it is no surprise that animals allocate time to monitoring potential competitors with which they can directly interact over food. Vigilance associated with direct food competition, which is also known as contest competition, can take place at two different stages during foraging: (1) when searching for resources and (2) when exploiting a food patch or handling food items.

2.3.1.1 Vigilance During the Search Phase

At first sight, vigilance directed at competitors during the search phase seems paradoxical because animals are not currently exploiting any resources. Nevertheless, a role for vigilance at this early stage has been identified in two contexts.

First, foragers in areas of low food density could use vigilance to glean information from conspecifics about the location of alternative food patches (Krebs, 1974; Lazarus, 1978; Waite, 1981). Of course, watching competitors for opportunities to increase foraging success could simply be called food searching rather than vigilance. However, the foraging behaviour of competitors can have a direct impact on survival by decreasing the relative share of limited resources available to an individual, and as such competitors are a threat to survival and what they do is worth monitoring.

This theory applies to species that forage on heterogeneous resources and that can keep a close watch on the foraging activity of nearby foragers. It predicts a negative relationship between vigilance and current food patch quality. This theory has received little attention and is rarely cited nowadays. However, it has inspired producer–scrounger (PS) models that, by contrast, have been developed extensively during the last 30 years.

The PS system represents the second context in which vigilance can be used during the search phase. In a PS system, foragers use the producer mode to locate resources on their own and the scrounger mode to locate food patches discovered by others (Barnard and Sibly, 1981). The scrounger mode resembles the vigilance tactic of unsuccessful foragers described earlier. However, the major distinction is that scrounging is expected to occur regardless of current success. Indeed, the time allocated to scrounging is predetermined within an individual and optimally set to fit the distribution of resources and the number of foragers in the habitat. In scrounger mode, foragers use vigilance to locate food patches discovered by others. Individuals can use producer and scrounger when searching for resources, but not at the same time. Therefore, an increase in the use of scrounger in a population will lead to a decrease in the overall rate at which resources are discovered.

Scrounging is expected to be more prevalent when food patches are relatively large, as it allows scroungers to obtain a greater share of the resources, and when foraging occurs in large groups since the corporate rate of food finding is likely to be higher (Giraldeau and Caraco, 2000). These are just some of the conditions where we would expect vigilance aimed at competitors to increase in a PS system. Recent PS models also showed that scrounging increases when food patches are more scattered (Afshar and Giraldeau, 2014; Beauchamp, 2008a).

In the earlier discussion, foragers use scrounger to locate food patches containing several food items that can be shared by all patch occupants. However, scrounging can also be invoked for single food items, as was the case for oystercatchers and their mussels. In this case, a prey item must be large enough to make scrounging, or food theft to be more precise, economically worthwhile. In northwestern crows foraging on beaches in western North America, extensive visual monitoring by a forager when foraging predicted the subsequent occurrence of food thefts (Robinette and Ha, 2001), suggesting that finding opportunities to scrounge food items from nearby foragers constituted one of the functions of vigilance in this species. Mexican jays tended to target less dominant group members for scrounging (McCormack et al., 2007), again suggesting that foragers can use vigilance to identify potential scrounging targets. A similar mechanism also operates in troops of brown capuchin monkeys (Hirsch, 2002).

Two issues in a PS system are relevant to a discussion of vigilance in the context of food competition. The first issue is how to practically distinguish vigilance aimed at predators from vigilance aimed at finding foraging opportunities through scrounging. In northwestern crows, the two types of vigilance involved the same type of visual monitoring, and only the outcome of vigilance, say, a food theft, suggested the target of the earlier vigilance. By contrast, a specific type of vigilance appears to be associated with scrounging in nutmeg mannikins, a small avian granivorous species that over the years has become a model species for laboratory investigations of scrounging. In this species, holding the head up, which is the potential marker of vigilance, can occur while stationary or while hopping on the ground. It turns out that the use of a stationary bout with the head up correlated with predation risk (the distance to a refuge in this case) but not with the amount of scrounging performed by an individual. By contrast, more hopping with the head up only led to more scrounging in this species (Coolen and Giraldeau, 2003) (Fig. 2.2).

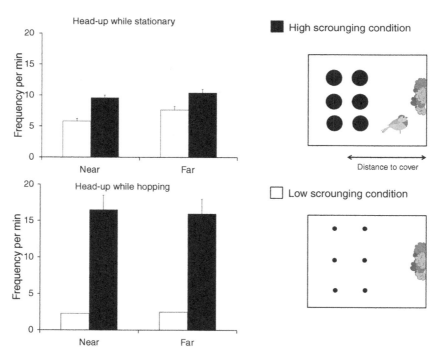

FIGURE 2.2   **A marker of vigilance to detect scrounging opportunities.** An increase in distance to cover, a correlate of predation risk, led to an increase in the use of stationary bouts with the head up in nutmeg mannikins but had no effect on the use of hopping with head up, suggesting that the former is used to monitor predators and the latter to detect scrounging opportunities. Hopping with the head up was more common with large patches that allowed more scrounging. A schematic version of the experimental set-up is shown on the right panel. Means and standard error bars are shown. *(Adapted from Coolen and Giraldeau (2003)).*

These results suggest that predation threats are assessed during a stationary bout with the head up while hopping with the head up facilitates scrounging (Wu and Giraldeau, 2005). Unfortunately, the results are not so clear cut; the strength of the association between vigilance postures and scrounging varied substantially from flock to flock (Coolen et al., 2001). In addition, even individuals that obtained most of their food from scrounging still hopped with the head down frequently. It is not easy, then, to tell which posture was used when scrounging. The search for a specific marker of vigilance aimed at competitors continues.

The second issue is whether vigilance aimed at predators is compatible with vigilance for scrounging purposes. It might be the case that investing in one type of vigilance interferes with the ability to detect threats of the other type. One PS model assumed that animals use the head-up posture to detect predators and scrounging opportunities at the same time (Ranta et al., 1998). This is an interesting development because it allows modelling scrounging in the context of predation, a feature that has eluded earlier models (Vickery et al., 1991).

At issue, then, is whether an animal can simultaneously carry two very different activities. As described earlier, laboratory experiments with nutmeg mannikins suggested that the two types of vigilance were carried out sequentially rather than simultaneously. However, recent laboratory evidence showed that zebra finches exposed to greater predation risk tended to be more vigilant and to use scrounging more often (Mathot and Giraldeau, 2008), suggesting that whichever posture is used for scrounging can be used to detect predation threats as well. Other studies also suggested a greater reliance on scrounging when predation risk increases (Barta et al., 2004; Bugnyar and Kotrschal, 2002). In all these studies, however, the mechanisms that produced an increase in vigilance might not have been exclusively related to scrounging. A robust answer to the incompatibility issue can only come from experiments designed to test the ability of individuals to detect predation threats, say, like the silhouette of predator, when the head is up during a stationary bout or while hopping.

Perhaps a third context during which vigilance can take place during the search phase has been identified recently. In many species of birds, nestlings compete with one another to gain access to resources brought to the nest by their parents. Nestlings that maintain a higher level of vigilance when the parents are absent might be the first one to react and receive food when a parent arrives at the nest. High vigilance for signs betraying the arrival of their parents was documented in barn owl chicks (Dreiss et al., 2013; Scriba et al., 2014). Vigilance in the nest might also serve the purpose of watching competitors to assess their position, information that would be useful when jostling for position to get fed. Nestlings are not foraging in the strict sense since they are dependent on an outside source of food. Nevertheless, vigilance by nestlings waiting to be fed resembles vigilance by a forager during the search phase as it can be used to increase future foraging gains.

2.3.1.2 Vigilance During the Exploitation Phase

Vigilance during the exploitation phase represents the flip side of vigilance during the search phase. When exploiting a patch, a forager can maintain vigilance to detect threats from nearby competitors. Food competitors could move in to share a patch amicably, as in the scrounging examples mentioned earlier, or use aggression to displace the patch owner or steal food items.

In one of the very first empirical papers examining the factors that influence vigilance in birds, Murton and his coworkers noted that wood pigeons in agricultural fields adopted an upright posture when stubble prevented them from easily seeing their surroundings (Murton and Isaacson, 1962). This posture also functioned as an alarm signal to others. They also noted that wood pigeons foraging at the edges of a flock looked wary, probably meaning that they used this upright posture more frequently (Murton et al., 1966). While it might be the case that vigilance was higher at the edges of a flock because peripheral individuals are first in line in case of an attack by a predator, Murton and coworkers suggested that peripheral birds directed their vigilance at dominant flock members located inside the flock that occasionally displaced them aggressively from food patches. When subordinate individuals act as food finders, as is the case in other species (Stahl et al., 2001; Tiebout, 1996), vigilance against threatening companions is probably a common tactic to avoid displacements.

Several studies have investigated the relationship between vigilance and the threat of food competition during the exploitation phase in birds and mammals (Table 2.1). In a particularly telling example, Eurasian oystercatchers that invested more in vigilance while handling large mussels were less likely to lose their prey to competitors (Goss-Custard et al., 1999), suggesting that part of the vigilance effort in these birds was allocated to monitoring competitors to reduce food losses. Overall, the prediction that vigilance increases in the context of food competition was generally but not always supported.

TABLE 2.1 Relationship Between Vigilance and Competition When Animals are Exploiting a Food Patch or Handling Food Items

Species	Observation	References
Birds		
Wood pigeon	Lesser competitors at the edges of groups were more vigilant	Murton et al. (1966)
Bald eagle	Vigilance increased with the threat of food theft	Knight and Knight (1986)
White-breasted nuthatch	Females were more vigilant than their more aggressive male partner	Waite (1987b)

TABLE 2.1 Relationship Between Vigilance and Competition When Animals are Exploiting a Food Patch or Handling Food Items *(cont.)*

Species	Observation	References
Tufted titmouse	Subordinates in foraging groups were more vigilant	Waite (1987a)
Willow tit	Dominant adults were more vigilant than subordinates	Hogstad (1988b)
Great tit	Vigilance unrelated to threats from conspecifics	Carrascal and Moreno (1992)
Eurasian oystercatcher	Vigilance during food-handling reduced the chances of losing food items to competitors	Goss-Custard et al. (1999)
Carolina chickadee	Subordinate individual in a pair showed higher vigilance	Pravosudov and Grubb (1999)
Serin	Vigilance increased in the presence of dominant males	Domènech and Senar (1999)
Bronze mannikin	No effect of food competition intensity on vigilance	Slotow and Coumi (2000)
Eurasian siskin	Vigilance increased with food competition	Pascual and Senar (2013)
Mammals		
Squirrel monkey	More time spent watching threatening group members than the non-social environment	Caine and Marra (1988)
White-faced capuchin	Females were less vigilant than males despite greater threat of food competition	Rose (1994)
Tasmanian devil	Vigilance aimed at competitors when foraging at carcasses	Jones (1998)
Yellow-footed rock-wallaby	Subordinates were more vigilant than dominant group members	Blumstein et al. (2001a)
Spotted hyaena	No relationship between social rank and vigilance	Pangle and Holekamp (2010b)
Blue monkey	Vigilance increased in the presence of dominant group members	Gaynor and Cords (2012)
Pine marten	Vigilance did not change in response to the presence of a food competitor	Wikenros et al. (2014)
Barnacle goose	Decrease in vigilance in larger flocks explained by more intense competition for food	Kurvers et al. (2014)

One difficulty with some of these studies is that predation risk might have varied inadvertently with the level of food competition, making it difficult to determine the target of vigilance. This was clearly a possibility in the wood pigeon study cited earlier in which the lesser competitors were also more exposed to predation at the edges of the flocks.

Whether vigilance is aimed at predators or at threatening food competitors is just as difficult to establish in the exploitation phase as it was during the search phase. In species with forward-facing eyes, the direction of the gaze can be used to identify the target of vigilance (Gaynor and Cords, 2012). The distinction between the two types of vigilance can also be facilitated when there is little overlap between the social and non-social environments. Favreau and his collaborators, for instance, focused on peripheral female Eastern grey kangaroos and determined that vigilance was aimed at conspecifics when individuals looked at companions inside the group, and aimed at predators, by contrast, when individuals looked away from the group (Favreau et al., 2010). Social vigilance was also invoked using a similar criterion in flocks of birds (Fernández-Juricic et al., 2005) and in bat colonies (Klose et al., 2009).

Food competition typically involves one foraging individual and nearby competitors eyeing their resources. Vigilance, in this case, allows the targeted forager to reduce the cost of competition and increase its own foraging success. In some cases, vigilance allows more than one individual to benefit from early detection; it can extend, for example, to nearby family members. Vigilance by one or many family members against potential trespassers, to the benefit of all family members, was noted quite early in the literature (Jenkins, 1944), and is especially common in waterfowl (Kotrschal et al., 1993; Lazarus and Inglis, 1978; Loonen et al., 1999). This type of vigilance can be viewed as a form of parental investment, a costly activity that increases future reproductive success and survival.

Circumstantial evidence also suggests that conspecifics represent a major target of vigilance even without direct food competition. This is the case in many primate species where intense monitoring of more dominant group members has been documented in non-foraging contexts (Haude et al., 1976; Keverne et al., 1978; McNelis and Boatright-Horowitz, 1998; Pannozzo et al., 2007; Treves, 2000; Watts, 1998). Knowing where dominant group members are and what they are doing is probably essential before engaging in any activity where competition matters. The fact that vigilance tends to increase when neighbours are closer represents another line of circumstantial evidence that monitoring competitors is important (Fuller et al., 2013; Hirsch, 2002; Kutsukake, 2006; Treves, 1999).

2.3.2 Scramble Competition for Food

Food competition has emerged as a major determinant of the level of vigilance maintained by animals. It influences vigilance when animals are searching for resources and more obviously during patch exploitation. In most of the cases cited earlier, food competition can be characterized as a contest (Sutherland, 1996) because losses follow direct interactions with other individuals. In contest competition, foragers use vigilance to detect competitors early to reduce food losses or avoid displacements.

Food competition can actually have a paradoxical effect on vigilance when competition involves speed rather than brawn. Competition of this type is referred to as scramble competition (Sutherland, 1996). Individuals must avoid predation and still obtain enough resources to survive. It is no surprise, then, that vigilance, which often takes time away from accumulating resources, should be sensitive to competition when the time to gather resources is of the essence. When resources are scare, any reduction in vigilance will allow an individual to obtain a disproportionate share of the contested resources (Clark and Mangel, 1986). Nearby foragers, in response, should also decrease their own vigilance to exploit resources just as quickly, bringing vigilance down for everyone. This is known as the milkshake effect, well-known to anyone who has ever shared a milkshake with unscrupulous friends.

Models suggested that in the face of scramble competition individuals are expected to decrease, rather than increase, their allocation of time to vigilance (Beauchamp and Ruxton, 2003; Bednekoff and Lima, 2004; Lima et al., 1999). Testing this prediction is tricky given that many factors can alter both vigilance and the level of competition at the same time. For instance, vigilance might be lower in larger groups due a perception that predation risk is lower and/or because the intensity of competition is higher in such groups.

One way to demonstrate an effect of scramble competition consists in showing how vigilance changes with the number of competitors when competition can be effectively ruled out, which could be done, say, by providing an effectively limitless amount of resources (Lima et al., 1999). Because animals could be hard-wired to compete by speed at all times, it is perhaps best to evaluate the effect of scramble competition on vigilance by actually manipulating the level of competition.

Few vigilance studies have directly manipulated competition intensity. In Tammar wallabies, the number of food bins available to a group was reduced to induce competition, but vigilance failed to decrease as predicted (Blumstein et al., 2002). In a field study with coots, competition was induced by providing a highly valued but scarce food (Randler, 2005a). Vigilance decreased when coots fed on the scarcer food, controlling for group size, suggesting an effect of competition on vigilance. A more recent approach simulated competition by adding virtual companions on a screen next to foraging birds in a cage (Rieucau and Giraldeau, 2009). Competition intensity was manipulated by varying the amount of time virtual companions spent feeding. When virtual companions appeared to feed more intensively, focal birds decreased their own vigilance to a greater extent, suggesting that perceived competition induced changes in vigilance (Fig. 2.3). An alternative explanation involves vigilance copying, a

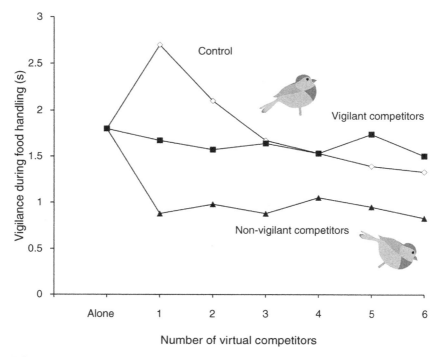

FIGURE 2.3 Scramble competition and vigilance. Nutmeg mannikins scanned less frequently on average when virtual competitors shown on a video-screen in their cages fed more intensely. The control condition consisted of virtual competitors normally alternating between head-up (vigilance) and head-down (feeding) postures. In the vigilant condition, virtual competitors only displayed the head-up posture. In the non-vigilant condition, by contrast, competitors always maintained the head-down posture. Errors bars are not shown for clarity. *(Adapted from Rieucau and Giraldeau (2009)).*

mechanism that will be covered in Chapter 7. Despite the admittedly limited support for an effect of scramble competition on vigilance, this mechanism is important in reminding us that competition can have varied and often drastically different effects on vigilance depending on the nature of interactions amongst competitors.

2.3.3 Competition for Mates

The previous section clearly established the importance of monitoring conspecifics in the context of food competition. It has long been appreciated that vigilance can also be aimed at conspecifics in the context of reproduction. Indeed, Jenkins noted early on that male snow geese during the reproductive season maintained a close watch to keep other males away from their mates (Jenkins, 1944). The fact that vigilance varies between the sexes, and often

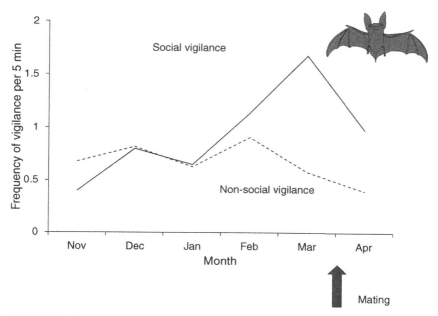

FIGURE 2.4   Vigilance fluctuations in relation to mating season. In Australian flying-fox colonies, the frequency of vigilance during focal observations varied seasonally mainly due to an increase in the social component of vigilance near mating season. *(Adapted from Klose et al. (2009)).*

follows seasonal patterns associated with reproductive status also strongly suggests mate competition (Fig. 2.4). As we will see below, vigilance can play various roles in the context of reproduction depending on the sex of the animals.

2.3.3.1   The Male Perspective

Males often compete with one another to gain or maintain access to females, and should have a strong incentive to monitor other males closely. Such vigilance fulfils various roles. In zebras, wildebeest and waterbuck in Africa, Burger and Gochfeld (1994) attributed high vigilance in territorial males to the detection of potential trespassers eyeing their females. In giraffes, bulls maintained more vigilance in groups than alone, suggesting that vigilance in groups was mostly directed at other group members (Cameron and Du Toit, 2005) (Fig. 2.5). In fact, the level of vigilance in bulls reflected group composition rather than the number of nearby conspecifics. Bulls were especially vigilant when their closest neighbour was a large bull. Large bulls use aggression to establish priority of access to reproductive females. Monitoring such threatening rivals could be used to avoid aggression. A similar reasoning can explain why vigilance in chimpanzee males increased in the presence of non-affiliative and thus potentially threatening males (Kutsukake, 2006).

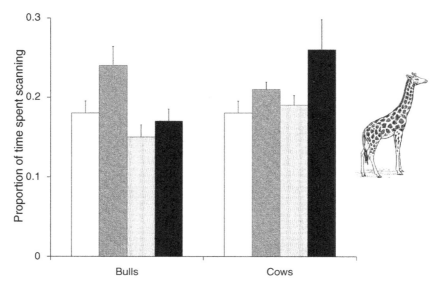

FIGURE 2.5   Mate competition and vigilance in giraffes. In bulls and cows, the average proportion of time spent scanning was higher in groups than alone. In groups, vigilance was highest when the nearest conspecific was a potentially threatening bull. Vigilance thus varied according to group composition: alone (open bars), with the same sex only (hatched bars), mixed sex (dotted bars), or with the other sex only (black bars). Means and standard error bars are shown. *(Adapted from Cameron and Du Toit (2005)).*

An increase in vigilance directed at rivals during the mating season can be viewed as a form of paternity insurance. Paternity insurance can be invoked as long as it is possible to distinguish social and non-social vigilance. In white-tailed ptarmigan, a small grouse species living in the tundra, paired males are extremely vigilant before the onset of incubation. High vigilance could serve the purpose of detecting rival males to ensure paternity in the fertile phase of the female. However, vigilance could also be directed at predators to allow the female partner to forage more safely at this particularly energetically demanding time of the year (Wittenberger, 1978). To distinguish between the two potential functions of vigilance, Artiss and Martin related vigilance patterns in paired males to correlates of predation risk and to correlates of the risk of paternity loss. It turns out that male vigilance tended to reflect the risk of predation rather than the risk of paternity loss, suggesting that while males are on the look-out for rival males one of the major functions of their vigilance remains to detect predators (Artiss and Martin, 1995). Males play a similar sentinel role in mated pairs of other species (Bull and Pamula, 1998).

The picture gets more complicated in species in which males are more colourful than females. Brighter males could be targeted disproportionately by visually oriented predators. Sexually dimorphic paired males might thus use vigilance to increase their own survival; higher male vigilance would only incidentally reduce predation risk for females. This potential role of male vigilance has been tested in species of ducks in which males tend to be more brightly coloured than females. In a strongly sexually dimorphic species, an unpaired male would be literally a sitting duck for visually oriented predators. In the Harlequin duck, a sexually dimorphic species, paired males displayed more vigilance than unpaired males, indicating that bright plumage was not the main cause of higher vigilance in males (Squires et al., 2007). As in white-tailed ptarmigan, correlates of predation risk explained the pattern of vigilance more successfully in paired Harlequin males. Higher male vigilance, it is conjectured, allows females to relax their own vigilance and focus more extensively on acquiring resources in risky situations. Other sexually dimorphic species also showed the same vigilance pattern (Guillemain et al., 2003; Portugal and Guillemain, 2011). These findings make one wonder whether females should not use the ability of males to maintain a high level of vigilance as a criterion for mate choice.

2.3.3.2   The Female Perspective

In the context of mate competition, females could use vigilance to detect or avoid threatening males. Harassment by males can cause a reduction in time spent feeding and also lead to injuries or even death (Li et al., 2012; Réale et al., 1996; Sundaresan et al., 2007). Vigilance could thus reduce such potential costs by giving females extra time to respond appropriately. Returning to giraffes, females, like males, showed more vigilance when closer to a bull than to another female (Fig. 2.5). This makes sense because bull giraffes harass females to assess their reproductive status, causing costly feeding disruptions (Cameron and Du Toit, 2005).

Female vigilance against aggressive males is also clearly a priority in some primate species in which males not only harass females but can even kill their offspring (Steenbeek et al., 1999; Treves, 1999). Females here use vigilance to avoid threatening males for their own sake and for the sake of their offspring. The role of vigilance in preventing infanticide has not been extensively investigated. A recent study with prairie dogs, a rodent species in which males are known to kill offspring, investigated the relationship between natural variation in infanticide risk and observed vigilance levels in mothers. Prevention of infanticide risk, however, did not appear to be an important determinant of vigilance in this species (Manno, 2007). Maternal watchfulness or parental vigilance in general is, of course, well-established in many species, and will be considered in Section 2.4, which focuses this time on the anti-predator function of vigilance.

2.3.4   Modelling Social Vigilance

The earlier discussion established that one of the main functions of vigilance is to monitor competitors. A further step in the functional analysis of social vigilance would be to predict the consequences of such vigilance. In particular, there is the question of how much time should be allocated to social vigilance. Time is typically considered a limited resource, and any increase in time spent vigilant must be at the detriment of time spent on other activities that contribute to the overall success of an individual in the struggle for existence. While many models predict how much time animals should allocate to anti-predator vigilance (see Section 2.4), there has been comparatively little effort to model social vigilance in an evolutionary context. Here, I describe the few models that focus on social vigilance.

One model examined the effect of scrounging on vigilance. Empirically, the monitoring effort allocated to scrounging cannot always be distinguished from vigilance aimed at predators. However, in a model, it is possible to examine how animals allocate time to monitoring conspecifics and predators, and determine how the two types of vigilance combine in groups of different sizes (Beauchamp, 2001). Without scrounging, anti-predator vigilance is expected to decrease with group size because individuals benefit from the monitoring effort of many and experience a reduction in predation risk by a simple dilution of the risk of individual attack (more on this in Chapter 5). By contrast, vigilance associated with scrounging is expected to increase in larger groups with more scrounging opportunities (Vickery et al., 1991). In the model, combining the two processes led to a pattern of almost no change in overall vigilance with group size.

Scrounging is but one of many factors thought to increase in larger groups. Aggressive interactions also tend to become more prevalent as group size increases (Caraco, 1979a), which should lead to more social vigilance in larger groups. The message, therefore, is that group-size dependent effects, such as scrounging and aggression, can counteract the expected decrease in anti-predator vigilance with group size. This conclusion reflects the bluntness of our measurement of vigilance, which often conflates different targets of vigilance. A recent study in Eastern grey kangaroos highlights the need to distinguish the various types of vigilance. The two types of vigilance can be distinguished in this species. As predicted by the model, overall vigilance failed to vary with group size because the decrease in anti-predator vigilance was compensated by an increase in social vigilance (Favreau et al., 2010).

Individuals in a group could monitor one particular aggressor or scan the entire group to detect potential attackers. By keeping tracks of such threats, an animal might be able to get advanced warning of impending attacks. Primate groups are often characterized by intense dominance struggles (Datta and Beauchamp, 1991), and monitoring neighbours can be clearly beneficial. Less known is whether such monitoring favours a particular spatial configuration of subordinate and dominant companions in a group. To address these issues, Evers et al. investigated the spatial structure in a group that would result from the simple physical avoidance of threatening companions without including any particular monitoring needs (Evers et al., 2012). In further models built on this avoidance model, the researchers then introduced the need to monitor a specific potential aggressor or the need to monitor the whole group to assess threats from many individuals. It turns out that adding particular monitoring needs increased the tendency of dominant companions to occupy central positions. The need to scan the whole group increased the spatial spread of the group.

This model leads to testable predictions about the socio-spatial structure of groups with dominance struggles. Interestingly, scrounging models also predict that scroungers should occupy central positions to get easier access to companions searching for food at the edges of the group (Barta et al., 1997). Because scroungers are often considered dominants, the two models make the same spatial prediction but from very different perspectives. Generally, the above models are important because they can show us not only who monitors whom, but also the possible consequences of monitoring conspecifics.

2.4   MONITORING PREDATORS

Galton speculated that the main function of vigilance in Damara cattle was a reduction in predation risk. Grazing with the head low in the grass reduces the ability of cattle to detect an approaching predator. Raising the head during vigilance allows individuals to see farther and detect predators before it is too late. While it seems obvious that vigilance plays a role in detecting predators, the fact that vigilance also serves other purposes, such as monitoring conspecifics, means that a careful assessment of vigilance is needed before concluding that vigilance is aimed at predators.

In the next sections I first discuss two general issues related to anti-predator vigilance: does it effectively allow early detection, and is it incompatible with other activities? Negative answers to these two questions would undermine basic assumptions about anti-predator vigilance. I will then proceed to a formal analysis of how much time should be allocated to anti-predator vigilance to maximize survival.

2.4.1   Early Detection

The simple expectation that vigilant animals can detect predators sooner has, surprisingly, received little attention. One approach to tackle this issue contrasts the ability to detect threats when animals are vigilant or not. In birds, visual vigilance is thought to occur when the head is up, The ability to detect a predator should thus be much reduced when the head is down. Kaby and Lind investigated the ability of blue tits, a small passerine bird, to detect an approaching flying model of one of their main predators. It turns out that blue tits feeding head down detected the model predator just as quickly as those in the allegedly vigilant head-up body posture, and also took the same time to escape (Kaby and Lind, 2003) (Fig. 2.6). Blue tits even escaped just as quickly when the model

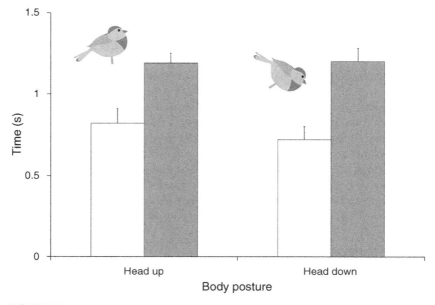

FIGURE 2.6 Body posture and predator detection. In blue tits, the body posture of an individual, whether head up or head down, had no impact on the mean time needed to detect a flying model of an approaching predator (open bars) or the time to escape (grey bars). Means and standard error bars are shown. *(Adapted from Kaby and Lind (2003)).*

predator approached from behind rather than the front. Obviously, our assumptions about predator detection ability in this species need to be reconsidered. Experiments of this kind are much needed to relate predator detection to body postures.

Another approach relates vigilance at the time of attack to the probability of escaping the approaching predator. The reasoning here is that if vigilance allows early detection, vigilant individuals should escape more quickly and be caught less often. Two mechanisms could explain why vigilant individuals survive better. First, predators might initially target less vigilant individuals (FitzGibbon, 1989). Second, the slow escape of less vigilant foragers might increase their chances of being captured. While this was not the case in the blue tit study above, fish (Krause and Godin, 1996) and other bird species (Elgar, 1986b; Hilton et al., 1999a) tended to escape more slowly when foraging head down.

When animals forage in groups, not all individuals are expected to be vigilant at the same time. Vigilant individuals at the time of attack should be able to escape first. Their hurried departures could provide a signal for unaware group members allowing them to flee after a delay (Lima, 1990). By analogy with the detection of predators, the vigilant body posture could also be useful to detect conspecific signals and initiate a speedier escape. A detection advantage to

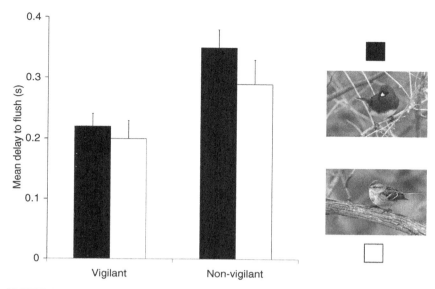

FIGURE 2.7   Detection advantage and body posture. In two flocking species of passerine birds, the dark-eyed junco (dark bars) and the American tree sparrow (open bars), the mean delay to flush to safety (s) after the rapid departure of a nearby conspecific after a mock attack was shorter when the non-detecting forager was vigilant rather than non-vigilant. Means and standard deviation bars are shown. *(Adapter from Lima (1994). Photo credits: Kelly Azar).*

vigilant animals within a group was addressed experimentally in a particularly convincing study by Lima. Lima used small balls rolled down a chute towards a foraging flock of birds below as a proxy for predation attempts (Lima, 1994). The ball could be aimed at one specific bird in the flock. Targeted birds flushed to safety as if they were attacked by a real predator. Lima noted how much time foragers next to a targeted bird took to flush to safety after a mock attack. In two species of birds, vigilant individuals next to the targeted bird took less time to flush than those foraging head down (Fig. 2.7). The difference in escape time between the two classes of birds was small, about 0.1 s, but would give a fast-swooping bird of prey more time to approach an unaware bird. This detection advantage within the group acts at a much shorter range than the typical distance at which a predator is detected, making a threat signal that much more conspicuous to the vigilant forager.

At any given time, the odds that one individual in the group is vigilant should increase with group size, as long as individuals monitor their surroundings independently from one another. When this is true, the ability to detect a predator at the group level should increase with group size. The earliest research on this hypothesis was performed with birds. In a classic study, a trained goshawk was used to attack flocks of wood pigeons from a standard distance (Kenward, 1978). The detection distance, as predicted, increased in groups with more pigeons. Similar results have been obtained since in many other species of birds (Boland, 2003; Cresswell, 1994; Harkin et al., 2000; Lazarus, 1979a; Powell, 1974; Siegfried and Underhill, 1975) and mammals (Da Silva and Terhune, 1988; FitzGibbon, 1990; Hoogland, 1981; Jarman and Wright, 1993; van Schaik et al., 1983). This line of evidence suggests that vigilance is useful to detect predators early.

Overall, the above evidence supports the view that vigilance provides early warning of attack. Nevertheless, it is clear that we need to know more about the sensory basis of predator detection before we can safely relate vigilance postures to detection ability.

2.4.2 Incompatibility with Other Activities

Vigilance was initially thought to be entirely incompatible with the acquisition of resources. This is because vigilance and other activities, such as foraging, are often carried out with postures only suited to one activity at a time. This was clearly the case for visual vigilance in the Damara cattle. Incompatibility has also been suggested for auditory vigilance. Noises produced during food handling can mask auditory cues associated with an approaching predator (Lynch et al., 2015), which would explain why animals periodically interrupt food handling to monitor surrounding noises.

Why does incompatibility matter? Without some level of incompatibility, there is no trade-off between vigilance and other activities. An animal could thus select the obvious option of being vigilant at all times to reduce predation risk. However, it seems very unlikely that vigilance could be maintained over a long period of time without any consequences for the ability to detect threat signals (Dimond and Lazarus, 1974).

The incompatibility between vigilance and other activities has probably been exaggerated. Recent research showed that vigilance and alternative activities are not necessarily mutually exclusive. Indeed, animals are able to maintain some vigilance while asleep (Rattenborg et al., 1999b; Ridgway et al., 2006) and when feeding head down (Bednekoff and Lima, 2005; Devereux et al., 2006; Lima and Bednekoff, 1999a). Mammalian herbivores can maintain vigilance while handling food (Cowlishaw et al., 2004; Fortin et al., 2004b; Makowska and Kramer, 2007). This is also true in many species of primates that feed in an upright position (Treves, 2000). However, it remains to be shown whether vigilance is just as effective when carried out alone or in combination with food handling or other activities. Interference between foraging and vigilance tasks carried out simultaneously has been documented in some species (Bohorquez-Herrera et al., 2013; Dukas and Kamil, 2000), ultimately reflecting limited brain processing ability (Marois and Ivanoff, 2005).

The trade-off between vigilance and foraging is not expected to be costly when animals can overlap different processes while foraging (Baker et al., 2011; Fortin et al., 2004a; Illius and FitzGibbon, 1994). When faced with easy foraging tasks, animals probably perform vigilance at little or no cost (Lawrence, 1985; Metcalfe, 1984; Moreno and Carrascal, 1991; Teichroeb and Sicotte, 2012). These results suggest that the degree of incompatibility between vigilance and other activities is not as extreme as first thought. The true cost of vigilance probably varies from species to species, although it seems unlikely that vigilance is typically cost-free. The occurrence of a trade-off between vigilance and other activities opens the door for formal analyses aimed at predicting how animals should split their limited time between vigilance and competing activities.

2.4.3   Modelling Anti-Predator Vigilance

The functional approach in animal behaviour research aims to uncover how a particular behavioural pattern influences survival or reproductive success. If natural selection shaped animal vigilance patterns in the evolutionary past, we should be able to predict exactly how much time individuals should allocate to vigilance in different contexts.

The above discussion showed that vigilance can reduce predation risk by providing early warning. The level of vigilance maintained by an animal should thus have quantifiable consequences for survival. Models are typically concerned with the survival consequences of vigilance although it is clear that vigilance can also influence reproductive success. Modelling anti-predator vigilance would be futile if vigilance failed to interfere with other fitness-enhancing activities like resting and foraging. Models are typically concerned with the trade-off between vigilance and resource acquisition. In a nutshell, an increase in vigilance means that there is less time for foraging because the two activities interfere with one another. Vigilance contributes to the fitness of an individual by reducing predation risk. Foraging allows an individual to accumulate resources, which are important to sustain life. The two activities ultimately contribute to fitness by increasing survival but in different ways (Box 2.1).

Models of anti-predator vigilance pit fitness against the level of vigilance adopted by a forager. No study that I am aware of has ever come close to documenting this fitness function for several reasons. The first challenge would be to sample individuals displaying different levels of vigilance in a habitat characterized by a fixed predation risk and a fixed rate of food intake. It is hard to imagine why individuals would stray too much from the optimal solution for the very simple reason that adopting a sub-optimal level of vigilance can be deadly.

Another challenge would be to convert a particular vigilance level into fitness units. Predation events are rare; to document the odds of surviving associated with a particular level of vigilance would require a lot of time and many study subjects. The fitness gains from foraging are also difficult to obtain as foraging only produces tiny increments in energy gain per unit time. Again, it would take a long time to see the reproductive consequences of foraging at a fixed rate (Lemon, 1992).

BOX 2.1 A Model of Anti-Predator Vigilance

I develop a simple model of anti-predator vigilance for a single forager that splits its time between foraging and vigilance. The two activities are considered mutually exclusive. During time T, the animal gains energy through foraging and is exposed to predation. The fitness of the individual, G, represents the product of two entities: (1) surviving predation through vigilance, and (2) the fitness gains obtained through foraging during time T. Fitness units could be the expected reproductive success over the lifetime of an individual. Natural selection should favour vigilance patterns that maximize fitness.

The animal is vigilant for a proportion v of its time and forages during the remaining proportion $1 - v$. While foraging, the animal gains energy at rate f. The foraging gains over period T are given by $e = f(1 - v)T$. These gains are transformed into fitness units with the function $F(e(v))$. Mortality through predation occurs only when the animal is foraging. The probability of failing to detect an attack while foraging can be modelled as $(1 - v)^2$ following Lima (1990). The probability of surviving all attacks is proportional to the attack rate (α), the time exposed to these attacks (T), and the probability of dying per attack. When attacks occur randomly according to a Poisson process, the probability of surviving attacks is given by: $\exp(-\alpha T(1 - v)^2)$. Therefore, fitness G is given by: $F \exp(-\alpha T(1 - v)^2)$.

Fitness G first increases with an increase in vigilance as the gains in surviving attacks more than compensate for the losses in foraging time (Fig. 2.8). However, such losses become more prominent as vigilance increases and the smaller expected gains in safety can no longer compensate. There is, therefore, an optimal value for vigilance that maximizes fitness. Simple algebra provides this optimal investment in vigilance:

$$v* = 1 - \frac{f(\partial F / \partial e)}{2\alpha F}$$

which indicates that the optimal vigilance increases with attack rate and decreases with the rate of energy gain while foraging.

This simple model illustrates the trade-off between foraging and vigilance, and how an optimal pattern of vigilance strikes a balance between two competing activities. This optimal value is a function of predation rate and the rate of energy gain. The model could be expanded by considering groups of animals foraging together. In this case, the probability of dying per attack becomes a function of group size. The relationship between vigilance and group size is investigated further in Chapter 5.

Faced with these nearly insurmountable challenges, we can only make progress by making simplifying assumptions about the relationship between vigilance and survival (Box 2.1). With reasonable assumptions, it becomes possible to predict adjustments in vigilance to a host of ecological factors that affect predation risk and the rate of energy gain while foraging. These factors will be reviewed in Chapter 4. The above caveats certainly remind us that the path from one wary scan to progeny years later is most convoluted.

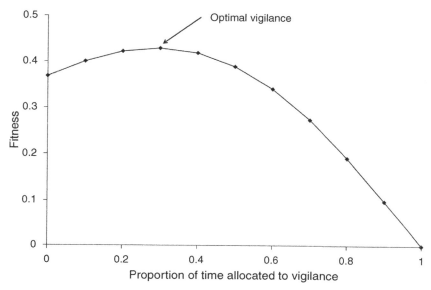

FIGURE 2.8   The trade-off between vigilance and foraging. As the allocation of time to vigilance increases, fitness first increases and then decreases, suggesting that losses in food intake are not recouped by improved predator detection. An optimal value for vigilance maximises fitness, which is the product of the probability of survival through a reduction in predation risk and the probability of survival through an increase in energy gains.

2.4.4   Temporal Organization of Anti-Predator Vigilance

Models of anti-predator vigilance, such as the one presented in Box 2.1, predict at best how much time should be allocated to vigilance. These models fail to specify how a given level of vigilance is practically achieved. The same level of vigilance over a given time period could be achieved in many ways: an animal could keep vigilance bouts short and alternate frequently between foraging and vigilance or spend more time vigilant each time and alternate less frequently (McVean and Haddlesey, 1980) (Fig. 2.9).

Our knowledge of the factors that influence the temporal organization of vigilance bouts is scant. A scan that is too short could miss crucial information, but too long a scan would probably yield redundant information (McNamara and Houston, 1992). The duration of an inter-scan interval probably reflects the ability of the forager to detect both its prey and its predators. For example, the ability to detect prey items might be more difficult if the foraging bout is interrupted too early or too frequently. This could be a problem for prey items that are difficult to detect against the background (Fritz et al., 2002). As for predator detection, early interruption of a foraging bout to initiate vigilance provides little benefit when predation risk changes slowly over time. However, it would be costly to sustain long foraging bouts if a predator can move in

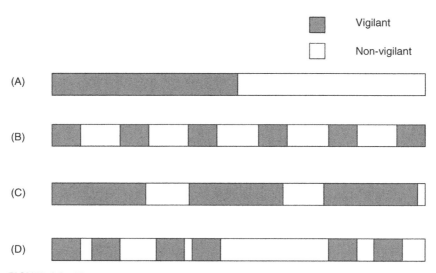

FIGURE 2.9 How to practically achieve vigilance. A forager can alternate bouts of vigilance with bouts of foraging in many different ways and still achieve the same overall level of vigilance. In all cases, time spent vigilant takes up 50% of the available time. In cases (A), (B), and (C), the forager initiates a bout of vigilance after a fixed amount of time spent foraging. When bouts of vigilance are more frequent, as in case (B), the forager spends less time vigilant each time. In case (D), the forager initiates a bout of vigilance after a random amount of time spent foraging, making vigilance bouts less predictable over time.

rapidly between scans (Sirot and Pays, 2011). Morphological features can also influence the expected pattern of vigilance. Foragers with a wider field of view, for instance, might not need to scan as frequently to monitor threats, suggesting that differences amongst species in visual fields could explain part of the variation in the temporal organization of anti-predator vigilance (Fernández-Juricic et al., 2010, 2011b; Moore et al., 2013; Tisdale and Fernández-Juricic, 2009). Without a detailed understanding of the costs and benefits associated with the length of a bout of vigilance or foraging, it is difficult to predict how vigilance should be organized over time.

Another issue altogether concerns the temporal variability of vigilance patterns. Should bouts of vigilance be initiated at regular intervals or more randomly over time (Fig. 2.9)? This issue is important because some vigilance patterns probably work best for particular types of predators. When vigilance is viewed as the probability of detecting biologically meaningful signals from the environment, the questions comes naturally as to how best to split the time between vigilance and any activity that interferes with detection. This question attracted the attention of operations researchers long before animal behaviour students. Operations researchers focus on the optimal way human subjects space vigilance bouts over time to detect objects that arrive in the field of view at unpredictable times, such as airplanes on a radar screen (Blackman and Proschau, 1957). The task of an air traffic controller is eerily similar to that of a forager aiming to detect predators that can attack at any time. Operations research concluded that evenly spaced vigilance bouts work best if the arrival of significant events cannot be predicted. We can now tell children in the playground to raise their heads from the sandbox at regular intervals to watch for the impending arrival of the ice-cream truck, which would strike a right balance between the enjoyment of playing and the early detection of the vendor.

In the very first model of predator detection, Pulliam (1973) assumed that predators attacked at random times but imposed an uneven initiation of vigilance bouts, presumably for mathematical convenience (Bednekoff and Lima, 1998a). In particular, Pulliam assumed that prey animals interrupt foraging to monitor their surroundings at random times. In the model, the distribution of foraging bout lengths, during which no vigilance occurs, follows a negative exponential distribution in which longer intervals are progressively less likely. This sharply contrasts with the even pattern predicted by operations research.

Subsequent models eventually showed that a regular pattern of vigilance works best when predators attack at random times (Bednekoff and Lima, 2002). In fact, an irregular pattern of vigilance over time works best only against observant predators (Scannell et al., 2001). Observant predators, such as lions in ambush, could use any regularity in vigilance patterns to launch their attacks when prey animals are expected to be least vigilant. Unpredictable initiation of vigilance in a foraging bout allows prey animals to evade observant predators.

In many species of birds and mammals, the distribution of inter-scan intervals poorly fits the negative exponential distribution, which is expected from Pulliam's model (Beauchamp, 2006; Carro and Fernandez, 2009; Pays et al., 2010; Ross and Deeming, 1998; Sullivan, 1985). Many of these species face non-observant predators, and a more regular pattern of vigilance is probably best-suited to locate potential sources of disturbance.

2.5   ADVERTISING TO PREDATORS

While anti-predator vigilance is typically viewed as an early detection system, it could, unexpectedly, also play a role in avoiding predation in the first place. The level of vigilance maintained by a prey animal could be advertised through a display aimed at predators, which would reduce the subsequent risk of attack. This is how it might work.

Some displays made by prey animals appear to inform predators about their condition. An impressive jumping display by a prey animal, for instance, can discourage the approaching predator from initiating a most likely futile chase against too vigorous a prey (Caro, 1986; FitzGibbon and Fanshawe, 1988). Signals made by prey animals, however, are often performed in the absence of immediate threats. Vacuous signalling might simply reflect the misidentification of threat signals. Misfiring, in this case, serves no purpose and only betrays general wariness. An alternative explanation, however, is that vacuous signals are performed dishonestly just in case a predator appears without being detected (Murphy, 2007). As such, these signals would act as pre-emptive strikes against newly arrived predators. Cheating of this kind by prey animals is thought to work as long as the signal is cheap to produce and not too common in the population.

Researchers have also argued that some displays produced in the absence of immediate threats could signal the level of vigilance maintained by a prey animal. Tail-wagging in the aptly named wagtail appears to serve this purpose: birds wag their tails more frequently when vigilant (Randler, 2006b). In California ground squirrels, tail flagging also appears to work as a signal of vigilance (Putman and Clark, 2015). Squirrels previously exposed to a predator snake flagged their tails more often and, crucially, reacted more quickly to simulated attacks, indicating a greater responsiveness to threats during tail flagging. The correlation between the intensity of tail flagging and reaction time indicates that this display reflects more than the false advertisement of detection; it effectively signals to the predator the level of responsiveness against threats.

What would prevent animals, such as wagtails and ground squirrels, from wagging their tails at all times, vigilant or not? These signals, for one, could be costly to perform, which would ensure honesty evolutionarily speaking. Production of these signals might reduce foraging efficiency, say, or attract the attention of other predators passing by. Production costs, nevertheless, are little known for vigilance signals aimed at predators.

A final piece of the puzzle still remains: would a predator avoid or be less likely to attack prey animals that signal their vigilance? The above findings suggest that the fear of attack can modulate the expression of vigilance and might be used to reduce the likelihood of future attacks. More work is needed to evaluate the consequences of producing these signals.

2.6   CONCLUSIONS

The early detection of threats, be they related to predators or conspecifics, represents the major function of vigilance. Giving animals a head start, vigilance allows individuals to escape predators earlier or detect and avoid threatening conspecifics. The adaptive framework generates predictions about the consequences of vigilance in the context of food or mate competition. In particular, social vigilance models predict the occurrence of subordinates at the edges of groups and an increase in the occurrence of social vigilance when animals scrounge the food discoveries of companions. Typically, the allocation of time to vigilance detracts from the ability to obtain resources, and adaptive models of anti-predator vigilance predict how much time prey animals should spend on vigilance in response to variation in predation risk and foraging contingencies.

While much is known about the trade-off between foraging and anti-predator vigilance, very little progress has been made regarding the optimal allocation of time to social vigilance. This probably stems from the fact that gains from social vigilance are difficult to convert into familiar units such as foraging rate or survival. A notable exception is when social vigilance serves to detect foraging opportunities, in which case vigilance can be modelled using the established PS framework. A further issue that requires more attention is exactly how to monitor potential aggressors or conspecifics to locate food patches. Regular or irregular anti-predator vigilance has been linked to predator type, and it is conceivable that there is an optimal way of scanning for different types of competitors.

Time and again the difficulty in distinguishing between potential targets of vigilance surfaces in the literature. This is an important issue because patterns of vigilance may be specifically tailored to a particular target. For instance, anti-predator vigilance is expected to decrease with group size while social vigilance should increase. Failing to distinguish between the targets of vigilance decreases the explanatory power of our models.

More knowledge about what the senses can perceive will become essential to determine the target of vigilance. Proximate issues regarding vigilance are covered in the following chapter. In this context, recent technological advances show promise for future research on animal vigilance. Breakthroughs involve the use of small head-mounted cameras that allow us to see what the animals see (Kane and Zamani, 2014) or head-mounted eye-tracking devices that can detect eye movements (Yorzinski and Platt, 2014). Video-tracking devices can be used to detect movements of the head during vigilance and measure their duration, speed and amplitude (Choy et al., 2012). Such technological developments could provide us with much finer measurements of animal vigilance.

Chapter 3

Causation, Development and Evolution of Animal Vigilance

3.1 INTRODUCTION

The previous chapter examined the possible functions of vigilance in animals, which include monitoring predators and competitors. The adaptive nature of vigilance is clear: it allows individuals to detect such threats early and react appropriately. The functional perspective can tell us how much time should be allocated to vigilance when animals face time or energy constraints. However, it remains silent on how vigilance is achieved at the proximate level, how it may vary during the lifetime of an individual, or how it evolved in response to past selection pressures. This chapter deals with the causation, development and evolution of animal vigilance.

3.2 CAUSATION

A proximate explanation seeks to identify the causal elements associated with vigilance. Typically, proximate explanations of behaviour focus on internal factors, such as hormones, brain structure and sensory ability, which modulate the expression of vigilance. In the vigilance scheme presented in Chapter 1, these factors influence brain responsiveness and monitoring effort. Essentially, these factors affect how the brain responds to signs of danger and the type of reactions animals produce in response.

3.2.1 Hormonal Factors

As we saw in Chapter 1, vigilance is often viewed as a state that allows the animal to detect important environmental stimuli. Hormonal factors play a role in vigilance by affecting the state of an individual. Hormones can influence the expression of vigilance at different levels. At the coarsest level, an animal needs to be awake to be fully vigilant, and hormones that control the sleep–wake cycle are thus ultimately involved in the control of vigilance (Dijk and Lockley, 2002). In some species of animals, in which it is difficult to identify a marker of vigilance, the wake state has in fact been used to define periods of vigilance (Lanham and Bull, 2004). While a non-sleeping animal can be vigilant, the wake state is not necessarily indicative of vigilance because many other

Animal Vigilance. http://dx.doi.org/10.1016/B978-0-12-801983-2.00003-6

activities not related to vigilance are carried out while awake. Other hormones are more intimately related to the expression of vigilance. The role played by these hormones is detailed in the following sections.

3.2.1.1 Testosterone

In the grey partridge, a sexually dimorphic bird, hormones appear to modulate the expression of vigilance in the context of mate competition (Fusani et al., 1997). Partridges monitor their surroundings using an upright posture, and this type of vigilance is more common in males during both the reproductive and non-reproductive seasons. During the reproductive season, male vigilance probably plays a role in detecting rivals, and the sex difference in vigilance suggests that sex hormones may be involved. To test this hypothesis, the authors implanted some males with testosterone, a gonadal sex hormone associated with the expression of secondary sexual characters in males, and compared their vigilance to that of untreated males. Treated males maintained a higher vigilance than non-treated males both before and after the presentation of a threat (Fig. 3.1), suggesting an increase in both routine and induced vigilance. Exactly how testosterone allows individuals to maintain more vigilance is not clear, but some studies suggested that testosterone affects how persistently individuals search for specific stimuli (Andrew, 1978). In humans, the level of testosterone has been related to increased attention to threat signals (van Honk et al., 1999), suggesting a role during social vigilance. Testosterone might thus influence vigilance by increasing monitoring effort. It would be interesting to see whether the ability to detect threat signals, given the same monitoring effort, also increases after the testosterone treatment. The fact that the secretion of hormones like testosterone follows diurnal as well as seasonal rhythms in many species (Wingfield et al., 1990) opens up the possibility of tying short- and long-term vigilance patterns to circulating levels of hormones.

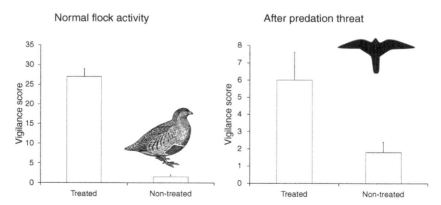

FIGURE 3.1 Testosterone and vigilance. Testosterone-implanted male grey partridges maintained a higher vigilance than non-treated males during normal flock activity and after the presentation of a raptor silhouette. Means and standard error bars are shown. *(Adapted from Fusani et al. (1997).)*

3.2.1.2 Oxytocin

Oxytocin has been linked to social vigilance in rhesus monkeys. Primates often live in very hierarchical societies, and monitoring other group members can provide crucial information about potential aggressors. Individuals that are adept at recognizing threats from conspecifics would benefit from early warning. However, given the time and energy costs involved, heightened social vigilance probably needs to be channelled to be effective.

As with the previous testosterone study, researchers used exogenous doses of oxytocin and examined consequences for social vigilance in captive rhesus monkeys (Ebitz et al., 2013). The results showed that while oxytocin increased attention to the eyes of other monkeys, it appeared to blunt social vigilance for threats (Fig. 3.2). Overall, an increase in oxytocin reduced the time and effort allocated to monitoring potentially threatening companions. By reducing social vigilance, oxytocin would favour prosocial behaviour (Chang and Platt, 2014), which fits with the idea that oxytocin, by contrast to testosterone, is a bonding hormone.

Individuals vary in the amount of oxytocin they produce, which might lead to observable differences amongst individuals in the level of social vigilance. It is also known that oxytocin levels vary with prosocial behaviour such as touch

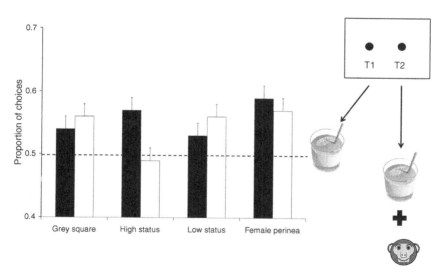

FIGURE 3.2   Oxytocin and social vigilance. Rhesus monkeys that inhaled oxytocin (white bars) rather than a saline solution (black bars) paid less attention to images of a dominant companion, suggesting blunting of social vigilance. In a pay-per-view experiment, monkeys faced a choice between two targets on a screen, one of which provided access to a juice reward and the other to the same juice reward and a picture. Four types of pictures were used: a grey square as a control, the face of a dominant individual, the face of a subordinate individual or the genital area of a female. The mean proportion of choices for the combination of pictures and juice is shown with standard error bars. Dashes show the random expectation of 0.5. *(Adapted from Ebitz et al. (2013).)*

or eye contact. In this case, we might predict a decrease in social vigilance after experiencing prosocial behaviour. The mechanism of action of this hormone involves brain structures like the amygdala, which is known to regulate attention and arousal to facial features in primates (Gothard et al., 2007). This structure probably plays an important role in social vigilance, and it would be interesting to investigate other hormones associated with this brain region.

3.2.1.3   Norepinephrine

When faced with danger, it would make sense for an animal to focus on the threatening situation and ignore non-crucial, rival sources of attention. The stress response in vertebrates allows just this. It consists of a series of short- and long-term physiological changes that let animals mount an effective defence against a potential threat (Charmandari et al., 2004). We have all experienced dangerous situations at one time or another: our heart rate increases, our senses are sharpened and we generally become more alert to danger. Such changes are driven by the stress response, which increases our ability to detect threat signals per unit of monitoring effort. This is how the stress response eventually connects with vigilance. As part of the stress response, animals release several hormones including norepinephrine. This hormone acts on the central and peripheral nervous systems, and is responsible for the activation of the sympathetic nervous system. Norepinephrine affects vigilance by focusing attention on the detection of threats.

One particularly interesting consequence of the release of norepinephrine is an increase in pupil size (Eldar et al., 2013). Changes in pupil size have been related to social vigilance in rhesus monkeys (Ebitz et al., 2014). With more dilated pupils, monkeys devoted more attention to socially relevant stimuli, which may be indicative of threats in the future. In peacocks, pupil size increased after the presentation of a predator model, again suggesting physiological arousal with measurable consequences at the pupil level (Yorzinski and Platt, 2014). While an increase in pupil size typically decreases visual acuity, it may also allow individuals to focus more on movement and high contrast features of the target, tuning out less relevant information during monitoring (Ebitz et al., 2014).

Measurement of pupil size typically requires an elaborate experimental setting, but changes in pupil size in some species can be documented without instruments (Feare and Craig, 1998). Pupil size could act as a marker of vigilance, and future work could help us understand how well it responds to social as well as predatory threats, and whether it predicts enhanced vigilance.

3.2.1.4   Cortisol

In response to an acute stressor, a neuroendocrine cascade takes place starting from the adrenal gland all the way to brain structures such as the hypothalamus and pituitary glands, the so-called HPA axis. This very quick reaction, often unleashed within minutes, releases glucocorticoids and other hormones in the blood stream helping the individual mount an adequate response to the stressor.

Exposure to predatory threats, for instance, produces spikes in cortisol, a so-called stress hormone released following activation of the HPA axis (Hawlena and Schmitz, 2010). Measuring the amount of circulating cortisol could thus provide useful information about the level of fear experienced by animals.

A recent study found that the amount of circulating cortisol decreased in sheep living in larger groups (Michelena et al., 2012). This was expected because the perception of predation risk, a major stressor for prey animals like sheep, is thought to decrease in larger groups. Concomitantly, the level of vigilance in this species decreased markedly with group size (Fig. 3.3), suggesting that cortisol plays a role in mediating the vigilance response perhaps by facilitating predator detection. However, the level of cortisol measured several times within the same individuals in groups of different sizes did not show a very strong correlation with group size. One problem with observational studies of this kind is that it is not possible to determine whether it is the increase in vigilance that causes changes in cortisol levels or the increase in cortisol levels that facilitates vigilance.

Sampling cortisol from the blood or the saliva, as was done with the sheep, is difficult for elusive or easily frightened animal species. Obviously, handling animals can cause stress and influence the very hormone we want to measure. For these reasons, cortisol and other hormones are often sampled from indirect sources like faeces. Corticosteroids, which include cortisol, measured from faeces collected opportunistically in the field, showed no relationship with vigilance levels in olive baboons (Tkaczynski et al., 2014), but tended to be higher in more vigilant meerkats (Voellmy et al., 2014).

Other studies adopted the experimental approach used earlier with partridges and rhesus monkeys. An experimental increase in corticosterone in male

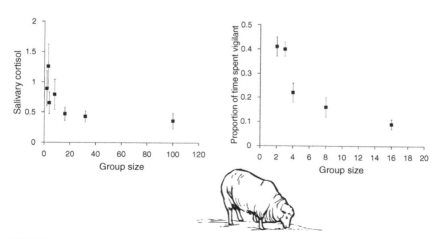

FIGURE 3.3   **Cortisol and vigilance in sheep.** Sheep in larger groups showed smaller levels of salivary cortisol (nanograms per millilitres) and less vigilance. Means and standard error bars are shown. *(Adapted from Michelena et al. (2012).)*

Adélie penguins failed to induce changes in vigilance (Thierry et al., 2014). In meerkats, an experimental increase in cortisol levels also failed to produce an increase in sentinel behaviour and in vigilance while foraging (Santema et al., 2013; Voellmy et al., 2014). In another study, barn owl chicks injected with corticosterone showed the same level of vigilance as untreated chicks (Dreiss et al., 2013). It might be the case that only a rapid increase in cortisol levels, or in other corticosteroids in general, in response to stress can cause a change in vigilance while persistently high levels associated with experimental treatment, as was the case in penguins, meerkats and barn owls, cause fewer changes. Overall, the link between corticosteroids and vigilance remains elusive.

Corticosteroids have been suspected to play a role in attentiveness to external stimuli and the consolidation of memory (Mateo, 2014). In Belding's ground squirrels, individuals with experimentally lowered levels of corticosteroids performed worse on a task involving the association of a warning call with appropriate defensive responses (Mateo, 2008). These results suggest that a spike in cortisol caused by a threat might facilitate an association between stimuli and appropriate responses. Since the baseline level of corticosteroids in juveniles can be high or low and tends to reflect maternal levels, the type of responses to threats shown by juveniles might be influenced by maternal effects, opening the door for indirect learning effects (Mateo, 2014).

The idea that fear sharpens the senses is not new. Darwin suggested long ago that emotions play an important role by facilitating the perception of danger and allowing a better response to the threat (Darwin, 1872). In the face of fear, the perception of the senses would be altered to increase detection ability, but exactly how this is achieved is not particularly well known.

Darwin noted that facial expressions in humans change remarkably with fear. Particularly striking are the widening of the eyes and the distention of the nostrils. Can such changes allow individuals to detect threats more easily? A recent study with human subjects mimicking different facial expressions showed that when expressing fear, individuals benefited from a wider visual field, moved their eyes faster during a targeting task, and inspired more air per breath (Susskind et al., 2008). All these changes would increase visual and olfactory perception of threats.

The ability to alter facial expression is more limited in other animals, but the possibility remains that sensory perception in the face of fear may be altered by other means. For instance, the adoption of bipedal vigilance in mammals increases the field of view and probably affects the ability to detect threats (Santema and Clutton-Brock, 2013). Movements of the ears can also help pinpoint the spatial position of a threat. Such changes represent a response to fear and must, therefore, reflect the secretion of fear-related hormones.

Vigilance is often viewed as a behaviour that alleviates fear in animals (Stankowich and Blumstein, 2005). This fear, a mental state, arises from potential threats caused by predators and conspecifics. Modulation of stress hormone levels, such as cortisol or norepinephrine, in response to such risk is consistent with an association between fear and vigilance in animals (Hawlena and Schmitz, 2010). Drugs like diazepam, which reduce anxiety levels in many vertebrate species, have been shown to reduce scanning behaviour in laboratory animals, as would be expected if vigilance reflects the level of fear experienced by an animal (Choy et al., 2012). Experimental injection of anti-anxiolytic drugs would probably cause a reduction in the ability to detect treats and evade predators. This seems a fascinating project for the future.

3.2.2   Neural Factors

The production of hormones is often intimately related to specific brain structures. In this way, vigilance involves brain circuitry needed for the production of hormones. In the next section, I show that specific adaptations in the brain also facilitate the detection of threatening stimuli during vigilance.

3.2.2.1   Neural Adaptations

Ethologists have long established that certain stimuli can instinctively elicit alarm in animals (Tinbergen, 1951). Ducklings with little exposure to predators nevertheless immediately react with fright upon sighting the silhouette of a predator flown above. This pre-conditioning is not limited to animals and extends to our species as well. Human subjects can detect a fear-relevant stimulus more quickly in a large set of images than fear-irrelevant stimuli (Öhman et al., 2001). In the same study, people afraid of snakes detected snakes more easily in a large set of distracting images. This agrees with the everyday experience that the best people to find spiders in a room are those who are afraid of them. These findings certainly suggest that some threat signals can capture attention more easily. But exactly how such targeted attention arises is not fully understood.

Returning to alert monkeys living in a strongly hierarchical society, recent research documented the existence of neurons particularly sensitive to stimuli that are important in a competitive social setting. Reviewing the physiological literature on specific triggers of neuronal activity in humans and primates, Emery reported the existence of neurons that are sensitive to eye gaze direction, face direction, body direction, and facial expressions (Emery, 2000). Given that signs of aggression in primates often involve gaze and body direction as well as facial expressions, the occurrence of such specialised cells provides a neural substrate to rapidly decode the visual information acquired during vigilance.

In birds, it appears that some species use specific morphological features of a predator, such as eye colour or beak shape, to classify species as dangerous (Beránková et al., 2014). Birds are also sensitive to a predator's gaze direction, which could be used to assess the level of interest of the predator (Freeberg et al., 2014). Gaze sensitivity appears to be widespread in animals (Davidson et al., 2014), which is not really surprising given that gaze cues are important in avoiding aggressive conspecifics and predators. It is not known whether specific cells in the avian brain fire more often when specific threat signals are perceived.

In rhesus monkeys, many of these specialized nerve cells occur in the superior temporal sulcus (STS) and the amygdala, a small almond-shaped structure located at the base of the brain. At least in this species, the STS appears important in recognising the eyes, head and body orientation, but it is the amygdala that attaches a socio-emotional significance to these signals (Emery, 2000). For example, the STS would be used to detect a stare by a neighbour. The amygdala, in turn, would recognize that this direct stare represents a sign of dominance, and elicit fear and avoidance behaviour in response.

In humans, the amygdala has been recognized as the centre responsible for the modulation of fear and anxiety (Sander et al., 2003). This structure acts as a fear module whose function is to activate psychophysiological reactions and emotions following the perception of threatening stimuli. Viewed as such, this structure must play an essential role in vigilance. At least in primates, lesions to the amygdala led to impaired judgement about facial features, which may be crucial to decode social threat signals. Because the amygdala is also involved in the association of neutral stimuli with fear, lesions in this structure should alter the ability to recognize threats and react appropriately. Lack of induction of fear makes the point of vigilance rather moot as perceived threatening stimuli would fail to activate an appropriate response. The view that is emerging now is that the amygdala indirectly monitors the environment for signals relevant to the organism, especially those linked to threats (Whalen, 1998). The amygdala might direct the senses to acquire further information about ambiguous signals.

This view of the function of the amygdala suggests that in a situation perceived as dangerous, attention should shift from a task-oriented mode, such as foraging, to a sensory-vigilance mode directed in part by the amygdala. Activation of the amygdala would enhance processing of sensory information to increase threat detection. This could be done by a direct route from the amygdala to the senses or by the secretion of hormone like norepinephrine, as we saw earlier. The end result would be a sharpening of the senses ideally suited to threat detection (Shackman et al., 2011).

While an increase in vigilance may reduce the risk of not responding to threatening stimuli, it may also come at the cost of responding too frequently to a large range of stimuli, some less threatening than others. In other words, heightened vigilance would increase sensitivity to risk but at the cost of specificity. A lack of specificity in reaction to stimuli would lead to false alarms. In my study species, the semipalmated sandpiper, a small North American shorebird, the birds face daily attacks by birds of prey such as falcons. When the tide covers their feeding grounds, sandpipers gather in large groups on the shore to roost. Sandpipers roost close to the cover that hides falcon attacks, which makes them especially vulnerable. If the above reasoning applies to birds, this acute risk of predation might favour a hypervigilant state where many signals, including those that are harmless, are perceived as dangerous. False alarms are especially numerous in roosting sandpipers (Beauchamp, 2010b), perhaps reflecting the decreased specificity of their alarm system.

The discussion thus far has focused on humans and non-human primates where lesions and their consequences have been documented and where brain imaging studies are feasible. Birds possess an equivalent of the mammalian amygdala in their brains (Zeier and Karten, 1971), and it would be expected that the above results also apply to them. Simpler organisms possess a smaller number of neurons, which would make them ideal subjects to investigate the neural basis of vigilance. However, there is virtually no research on vigilance in such species.

A recent study on medicinal leeches illustrates the potential for such research (Harley and Wagenaar, 2014). The researchers investigated the potential function of scanning behaviour in this species. While moving around, leeches frequently stop and move their heads and bodies in circular motions. Scanning movements in leeches are usually associated with the visual and/or tactile detection of prey items. However, their occurrence in the absence of any external stimuli suggests that scanning might also be used to detect predators. Scanning in leeches appears to be under the control of one neuron that actually interacts with the circuitry controlling the usual crawling motion. Studying the factors that affects firing in the motor neuron controlling scanning might help us understand how scanning is initiated in this particular species, and, more generally, the neural pathways involved in the control of vigilance.

3.2.2.2   Sleep

Sleep and vigilance are intimately related, opening yet another window to study the proximate control of vigilance in animals. While fully asleep, animals are considered very vulnerable to predators (Lima et al., 2005). As a consequence, the ability to structure sleep to maintain a sufficient level of vigilance must be under strong selection pressure (Scriba et al., 2014). How to maintain vigilance and sleep at the same time is at first sight considerably more difficult than being vigilant during foraging, as sleep temporarily shuts down the senses available for monitoring predators. One solution to this problem consists in interrupting sleep periodically to monitor threats (Amlaner and McFarland, 1981; Gauthier-Clerc et al., 1998; Lendrem, 1983b; Mathews et al., 2006). Depending on the rate of eye blinking, this strategy leaves individuals less protected when the eyes are closed.

The discovery of unihemispheric sleep represents one particularly interesting recent development in sleep research. This type of sleep is common in aquatic mammals and widespread in birds and other vertebrates like lizards. During unihemispheric sleep, one half of the brain is asleep while the other half remains awake to perform vital activities like keeping motor control for air-breathing aquatic mammals or detecting predators (Rattenborg et al., 2000). In unihemispheric sleep, animals can maintain a constant level of vigilance during sleep rather than the episodic vigilance associated with the eye-blinking strategy.

Asynchronous eye closure during sleep has been linked to unihemispheric sleep in animals (Rattenborg et al., 2000), providing a useful marker for this type of sleep. It allows individuals to monitor threats with one eye while resting the other half of the brain. If unihemispheric sleep serves an anti-predator function, it should be more prevalent where the risk of predation is higher. In addition, the open eye should face the direction from which predators are most likely to attack. These predictions were tested in flocks of mallards (Rattenborg et al., 1999a, b). The results showed that birds sleeping in the centre of the flock tended to use unihemispheric sleep less often than those at the edges, which are more exposed to predators attacking from the outside of the flock. Edge birds also oriented their open eye in the direction from which attacks by predators was most likely (Fig. 3.4). The open eye during unihemispheric sleep was also able to react quickly to visual stimuli. Mallards adaptively select the type of sleep that strikes the best compromise between safety and the need to rest the brain. I note that the strategy of using only one eye for vigilance may be ill-suited to habitats where predators can attack from many directions.

The neural circuitry and general physiology underlying unihemispheric sleep is poorly understood. A promising avenue may be to compare the neuro-anatomy of species with and without unihemispheric sleep to uncover structures

FIGURE 3.4   Unihemispheric sleep in mallards. Unilateral eye closure, indicative of unihemispheric sleep, prevailed at the edges of a flock of mallards. Mallards oriented their open eye more often away from the rest of the group than those at the centre. Dashes show the random expectation of no specific orientation. Potential threats in this laboratory setting originated from the outside of the group. Means and standard error bars are shown. *(Adapted from Rattenborg et al. (1999a).)*

that are involved in the control of this type of sleep. Such mechanisms are likely to play an important role as proximate determinants of vigilance.

3.2.2.3 Laterality

The naïve expectation has always been that the two eyes are equally effective in detecting visual threats. However, recent research indicates that since the two sides of the brain often process different types of information (Rogers and Andrew, 2002), the two eyes may not be equivalent when performing certain tasks. Tasks might thus be performed preferentially with one eye or the other. A left-eye bias appears to predominate when responding to predators (Austin and Rogers, 2012, 2014; Bonati and Csermely, 2010; Broder and Angeloni, 2014; Lippolis et al., 2002; Rogers et al., 2004) and aggressive companions (Austin and Rogers, 2012, 2014; Hews and Worthington, 2001) while the right eye seems to be used preferentially during foraging (Vallortigara et al., 1998). Nevertheless, there has been little work in the field (Austin and Rogers, 2012; Beauchamp, 2013a; Koboroff et al., 2008; Randler, 2005b; Ventolini et al., 2005), making it difficult to determine the extent to which preferences occur in the wild and their ecological relevance.

Most of the above studies documented a bias for the left eye when monitoring approaching predators. But is there any evidence for preferential eye use during pre-emptive anti-predator vigilance? In a semi-natural study addressing this issue, one avian species showed a preference to scan with the right eye while the other preferred the left eye (Franklin and Lima, 2001) (Fig. 3.5). The results are difficult to interpret because the preferred eye could not always be used to focus on the riskier side of the habitat. In another field study, one of three species of birds showed a preference to orient the right eye towards danger (Randler, 2005b). In my own field work with semipalmated sandpipers, birds achieved a higher foraging success when their right eye faced the direction from which most attacks by falcons originated (Beauchamp, 2013a), suggesting that a trade-off between foraging and vigilance is easier to achieve with the right eye. A similar result was obtained in fish that could monitor predators with one eye and search for food with the other (Dadda and Bisazza, 2006). In horses, anti-predator vigilance seems to be carried out with the left eye, but the direction of danger was not established (Austin and Rogers, 2012, 2014). It is important to establish the potential target of vigilance in such studies given that monitoring aggressive neighbours and predators could be performed preferentially with different eyes.

To understand the evolution of lateralization, as expressed by eye preferences, it is necessary to document the costs and benefits of using the preferred and less preferred eye. Achieving this goal will be difficult in the field because animals can compensate for inherent vulnerability. For instance, individuals might adjust their position in the group differently to focus their preferred eye on the riskier side of the habitat or live in groups to reduce their overall predation risk.

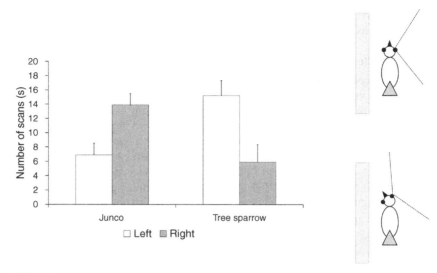

FIGURE 3.5 Lateralization of vigilance. When a visual barrier blocked their view on one side, dark-eyed juncos tended to scan in towards the obstruction more frequently when their right eye faced away from the obstruction while the opposite was true for American tree sparrows. The inset shows that by moving the head towards the obstruction, a bird with a right-eyed preference for anti-predator vigilance would increase its coverage of the area where predators are more likely to attack. *(Adapted from Franklin and Lima (2001).)*

The occurrence of brain lateralization can lead to preferential eye use, but such biases are rarely very strong. Few animals are expected to exclusively use one eye to locate predators or find food. The occurrence of partial preference for one eye during vigilance allows us, however, to examine how animals structure their vigilance when using the preferred and less preferred eye. For instance, it would be interesting to see whether animals compensate by scanning longer or more frequently when using the less preferred eye during vigilance. The discussion about brain lateralization and the occurrence of eye preferences during vigilance certainly highlights the importance of a better understanding of the proximate determinants of vigilance in animals.

3.2.3 Sensory Factors

Senses allow the detection of threats, and as such their structure and limits must surely constrain vigilance. Senses play a role in vigilance by modulating the effectiveness of the monitoring effort. Earlier research made simplifying assumptions about sensory perception during vigilance. Recall how researchers initially assumed that animals foraging head down would not be able to detect predators. Similarly, the detection of threats was assumed to be perfect when animals are vigilant. The recent surge of interest in proximate explanations of animal behaviour allows us to refine our assumptions about the senses. This has been particularly the case with vision, one of the main senses involved during vigilance. The following discussion thus relies heavily on recent findings from vision research. Research on avian vision has clearly established that the ability to detect important visual stimuli is intimately related to the configuration of the eyes and the arrangement of the photoreceptors on the retina.

I focus first on the configuration of the eyes. We tend to imagine how other animals see their worlds with our own binocular vision in mind. In many species, however, the eyes occupy a more lateral position, which widens the field of view on either side of the head. Laterally placed eyes in rats and many species of birds allow individuals to detect visual stimuli above their heads even when looking down (Lima and Bednekoff, 1999a; Tisdale and Fernández-Juricic, 2009; Wallace et al., 2013). In an ingenious experiment aimed at determining whether ground-foraging birds with laterally placed eyes value their peripheral vision when looking down to search for food, researchers blocked the side view of foragers with visual barriers (Bednekoff and Lima, 2005). Birds in the occlusion treatment clearly adjusted their scanning behaviour by looking up longer to compensate for the lack of visual information available when foraging head down (Fig. 3.6).

To investigate the consequences of eye configuration, it is also possible to compare patterns of vigilance in species with different fields of view. This is the approach taken in a recent study of two common backyard species of birds, the starling and the house sparrow. These two species differ markedly in their ability to monitor areas above their heads. The starling has a wider blind area at the rear of the head than the sparrow (Tisdale and Fernández-Juricic, 2009). Starlings appear to compensate for the wider blind area by increasing scanning bout duration to provide a wider coverage of their surroundings.

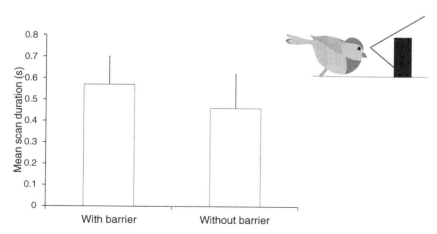

FIGURE 3.6   **The value of peripheral vigilance.** The presence of a visual barrier that prevented peripheral vigilance increased the mean duration of scans in dark-eyed juncos. Mean and standard deviation bars are shown. *(Adapted from Bednekoff and Lima (2005).)*

Eye configuration is expected to influence visual coverage. But the sharpness of detection is related to the density of photoreceptors in the retina, the structure involved in light detection (Martin, 2007). One retinal specialization in particular, the fovea, has important consequences for vigilance. The fovea is an area of the retina characterized by a high density of photoreceptors. High density provides a sharper visual resolution and may also enhance the detection of small movements, which are both crucial for visual predator detection. The arrangement of the fovea area and the number of foveae vary between species (Fernández-Juricic, 2012).

Areas of greater sensitivity in the retina matter because they can increase the value of information gathered during scanning. A prey animal could focus these more sensitive areas to particular parts of their habitats to detect threats more accurately. Some species can move their eyes to align sensitive areas to a region of interest. Species with relatively immobile eyes can orient the head in different directions during a vigilance bout to compensate.

Head movements of this kind during a vigilance bout have long been noted in the literature (Bekoff, 1995; Cézilly and Brun, 1989; Fernández-Juricic et al., 2011a; Jones et al., 2007; Lazarus, 1979a), but their function was not clear. Head movements during a vigilance bout can certainly increase the breadth of coverage by widening the field of view. Perhaps more importantly, head movements might be used to bring regions of interest to retinal specializations, such as the fovea, for better resolution. Head movements during vigilance would thus influence the quality of the information acquired during vigilance (Jones et al., 2007). If this is true, what the animal does while vigilant may be just as important as the time devoted to vigilance, which has always been the currency of choice to evaluate the effectiveness of vigilance.

Fernández-Juricic and I investigated head movements to assess their role during vigilance in cowbirds, a passerine bird better known for its habit of laying eggs in the nests of other species (Fernández-Juricic et al., 2011a). We found that peripheral individuals in a group, which are considered more vulnerable to predators, moved their heads more frequently than central companions probably to increase visual coverage of their surroundings. However, we found that in smaller groups and when individuals were generally farther apart, movements of the head decreased in frequency allowing individuals to fix their attention longer on specific features of their surroundings. In this case, individuals might be using the specialized parts of their retina to increase the quality of the information acquired during vigilance. Without eye-tracking devices, we could not readily identify the target of this sustained vigilance. Head movements during a vigilance bout certainly add a novel dimension to vigilance effort.

If areas of greater sensitivity in the retina play a role in vigilance, head or eye movements during vigilance should occur more frequently when the area of highest resolution in the eye is smaller, when the width of the blind area is wider, and when visual acuity is generally lower (Fernández-Juricic, 2012).

Testing such predictions will help us increase our understanding of the sensory basis of vigilance in animals.

The consensus that is now emerging from vision research is that retinal specialisations should be tuned to the specific visual challenges faced by animals in their habitats. As a case in point, consider animals foraging in a relatively open habitat in which features of interest tend to lay close to the horizon. In this type of habitat, predators can attack from all directions, but rarely from above or below, marking the horizon as a region of particular interest. A species with a small fovea and a small lateral field of view would need to make considerable head and eye movements to cover the whole horizon and would probably need to invest more time in vigilance than a species with a wider field of view and an extended area of increased visual resolution in the eye. Not surprisingly, many species living in open habitats, including birds, mammals and fish, possess horizontal visual streaks in their retina that allow them to see the horizon with greater resolution (Tyrrell et al., 2013). Close matching between features of the eyes and ecological challenges highlight the need for a better understanding of the role of sensory factors on animal vigilance.

3.2.4   Other Physiological Factors

When vigilance is viewed as the outcome of physiological processes, it becomes possible to understand and predict how internal and external factors can influence the ability to detect relevant stimuli in the environment. In this last section dealing with the proximate control of vigilance, I want to draw attention to a vast literature in humans that may be highly relevant to research on animal vigilance.

Queries on internet search engines using vigilance as a keyword are quite likely to return hits that belong to the human literature, which can be frustrating when we want to target animal vigilance research. Nevertheless, this body of work has many intriguing parallels with animal vigilance. The human literature on vigilance focuses on internal and external factors that affect performance on vigilance tasks, broadly defined as any monitoring task performed over a set amount of time that requires individuals to detect sporadic stimuli. A typical vigilance task in the laboratory involves detection of a stimulus that can appear anywhere on a computer screen at unpredictable times. The need to understand how people sustain attention to detect important stimuli drives research in this field. Findings from such research can be applied to everyday tasks such as paying attention to widely scattered road signs when driving or listening without snoozing to a long lecture on the proximate determinants of vigilance.

One particularly relevant finding from human vigilance research concerns the so-called vigilance decrement, namely, the decrease in the ability to perform well on a vigilance task as the time on task increases (Pattyn et al., 2008). This decrement has been variously attributed to boredom or to mental fatigue. In the case of boredom, the lack of stimulation during a task causes the mind to wander with a resulting loss in performance. In the case of fatigue, the decrease in performance arises from the strain of sustaining attention over a long period of time. Simply put, boredom arises because various attention-grabbing processes compete with one another while fatigue reflects the decrease in arousal when the same level of attention is maintained over a long period of time. Can boredom or fatigue be relevant to animal vigilance for predators or competitors? Here, I make a case for the involvement of these two factors in the context of anti-predator vigilance, but I suspect the same argument probably applies to social vigilance.

In many respects, vigilance against predators mirrors the play by Beckett in which the main characters wait forever for the famous Godot to arrive. On the one hand, a bout of vigilance against predators is typically futile since predation is a rare occurrence. On the other hand, maintaining vigilance is demanding since many possible stimuli must be processed repeatedly over time. Do animals get bored watching for rare predators or fatigued after a long watch?

The best way to answer this question would be to determine the ability to detect a predator-related cue as a function of the time spent vigilant up to that time. Both boredom and fatigue would predict a decrease in the ability to respond to such stimuli over time. Fatigue would predict an overall decrease in the ability to respond to many different types of stimuli, but this is not the case for boredom, which provides us with a way to distinguish between the two hypotheses. Sentinel behaviour is perhaps the best empirical system to investigate these questions. Sentinels in many species of birds and mammals maintain a vigilant state over extended periods of time (McGowan and Woolfenden, 1989), which allows us to evaluate the role of boredom and fatigue on animal vigilance. The consequences of a vigilance decrement in the context of anti-predator vigilance are fascinating. To give just one example, observant predators should preferentially attack later during a vigilance bout when prey vigilance is expected to be least effective.

Many exogenous factors have been shown to influence the ability to detect stimuli in human vigilance tasks. We are all familiar with the boosting effects of caffeine on alertness. Many other drugs have been shown to have a positive or negative effect on vigilance mostly through the modulation of neurotransmitters (Koelega, 1993), which are probably involved in the perception of threats. Such studies are useful since they can pinpoint regions of the brain involved in maintaining vigilance as well as identify neurotransmitters or hormones most likely involved in the vigilance process. It would be worthwhile extending this line of work to human vigilance tasks that simulate threatening rather than neutral or arbitrary stimuli to make it more relevant to animal vigilance research.

Also relevant to animal vigilance is the finding that sleepiness reduces vigilance performance in humans (Kribbs and Dinges, 1994). Animals are also expected to be sleep deprived at one time or another. For instance, the need to sleep can overcome other priorities after a long migration flight (Schwilch et al., 2002). Disturbances that interrupt normal sleep also probably increase sleepiness. If sleepiness influences animal vigilance, predation should be more successful when prey animals are sleepy and less likely to react optimally to predator-related cues.

3.3   DEVELOPMENT OF ANIMAL VIGILANCE

Acquisition of information from the environment through vigilance involves senses, neural processes, and a mobilisation of the body governed by muscles. Senses, nerves, and muscle fibres all result ultimately from the expression of genes, which direct the production of proteins in the different cell types involved. Like any behavioural pattern, one part of vigilance traces its origin down to genes.

Genes are necessarily involved in the production of behavioural patterns, but rarely sufficient. Some simple behavioural patterns can be expressed soon after birth with little input from the environment. Recently hatched ducklings with no experience of predators still react with fright when they see the silhouette of a predator flown above (Melzack et al., 1959). Most behavioural patterns, however, reflect a complex interaction between genes and the environment (Bateson, 1978). It is telling that even identical twins raised apart can show marked differences in behaviour despite sharing all their genes (Tellegen et al., 1988). An interaction between genes and the environment allows flexibility in the expression of behaviour amongst individuals. In the gene–environment interaction framework, it is of particular interest to determine how behaviour changes with age to pinpoint when and how the environment plays a role.

Two approaches have been used to investigate how vigilance changes with age in animals. In the first approach, perhaps the most powerful, changes in vigilance are investigated in the same individuals as they get older (Loughry, 1992). Longitudinal studies of this kind require long-term data on the same cohort of individuals, a daunting task for many mobile species, which may explain why this approach is rarely used. In a longitudinal study, each individual serves at its own control for future developments, a tremendous boost for statistical power. However, the results of a longitudinal study may be difficult to interpret when other influential variables also change during the time course of the study. Many variables can conceivably change as individuals get older including not only their experience, but also their size and physiological needs, all of which can influence vigilance and confound the effect of age.

The second approach, which is more commonly used, compares different individuals at different time periods. One potential drawback of this cross-sectional approach is that age might correlate with uncontrolled factors that have a direct bearing on vigilance. In tits, for instance, juveniles tend to forage in more exposed parts of the habitat (Gustafsson, 1988), a shift that would increase vigilance regardless of age. Another limit of this approach is that individuals with more extreme patterns of vigilance could be less likely to survive. If individuals with low vigilance at a young age die more frequently, say, vigilance will tend to increase with age, a trend that would not be apparent with the longitudinal approach. A further difficulty is that when comparing vigilance in different individuals, it is not always obvious which group shows normal vigilance. In a study in which juveniles appear less vigilant than adults, it might actually be the adults that are extra vigilant (McDonough and Loughry, 1995). Cross-sectional studies, just like longitudinal ones, are thus open to interpretation challenges.

Changes in vigilance with age have been documented in various contexts, including foraging and sleeping. Because few studies have examined age-related changes in vigilance while sleeping (Gauthier-Clerc et al., 2002; Shaffery et al., 1986), I focus here on vigilance during foraging. Does vigilance in this context vary as a function of age? To address this question, I compiled the available evidence using a broad survey of the vigilance literature in birds and mammals (Table 3.1). Not surprisingly, most of the studies adopted the cross-sectional approach. Overall, about two-thirds of the studies reported lower levels of vigilance in juveniles than adults (51/75, 68%). However, the pattern was far from universal, and the suspected underlying causes varied extensively.

TABLE 3.1 Effect of Age on the Amount of Time Allocated to Vigilance While Foraging in Birds and Mammals

Species	Type of study*	Direction of effect	Probable causes	References
Birds				
Yellow-eyed junco	L–C	Increases with age	Relied on parental vigilance	Sullivan (1988)
Giant Canada goose	L–C	No relationship with age; juveniles < adults	Parents provided safe foraging conditions	Seddon and Nudds (1994)
White-fronted goose	C	Juveniles < adults		Owen (1972)
Brent goose	C	Juveniles < adults		White-Robinson (1982)
White-winged chough	C	Juveniles < adults	Less efficient foraging by juveniles	Heinsohn (1987)
White-browed sparrow-weaver	C	Juveniles < adults	Territorial adults maintained more vigilance	Ferguson (1987)
Canada goose	C	Juveniles < adults		Austin (1990)

TABLE 3.1 Effect of Age on the Amount of Time Allocated to Vigilance While Foraging in Birds and Mammals *(cont.)*

Species	Type of study*	Direction of effect	Probable causes	References
Snow goose	C	Juveniles < adults	Less efficient foraging by juveniles	Bélanger and Bédard (1992)
Piping plover	C	Juveniles > adults		Burger (1991)
Common crane	C	Juveniles < adults	Less efficient foraging by juveniles	Alonso and Alonso (1993)
Willow tit	C	Juveniles = adults		Brotons et al. (2000)
Common crane	C	Juveniles < adults	Less efficient foraging by juveniles	Avilés (2003)
Siberian jay	C	Juveniles < adults	Parents provided safe foraging conditions	Griesser (2003)
Whooper swan	C	Juveniles < adults		Rees et al. (2005)
Tundra swan	C	Juveniles < adults	Relied on parental vigilance	Badzinski (2005)
Common crane	C	Juveniles < adults	Higher feeding requirements in juveniles	Avilés and Bednekoff (2007)
Greater flamingo	C	Juveniles < adults	Relied on parental vigilance to meet higher energetic demands	Boukhriss et al. (2007)
Black-necked crane	C	Juveniles < adults		Wang et al. (2009)
Snow and Ross's goose	C	Juveniles < adults	Less efficient foraging and competition with adults; higher feeding requirements of juveniles	Jónsson and Afton (2009)
Red-crowned crane	C	Juveniles < adults	Inexperience	Li et al. (2013)

(Continued)

TABLE 3.1 Effect of Age on the Amount of Time Allocated to Vigilance While Foraging in Birds and Mammals *(cont.)*

Species	Type of study*	Direction of effect	Probable causes	References
Mammals				
Dwarf mongoose	L	Juveniles < adults	Learned sentinel behaviour from adults	Rasa (1989)
Black-tailed prairie dog	L–C	Higher at first and then lower in juveniles	Vulnerability to predation and inexperience	Loughry (1992)
Yellow baboon	L	Negative relationship with age	Older juveniles spent less time monitoring mother	Alberts (1994)
Thirteen-lined ground squirrel	L–C	Juveniles < adults	Higher feeding requirements in juveniles	Arenz and Leger (1997a, 2000)
Moose	L	Juveniles = adults	More variation amongst juveniles reflected variation in body condition or inexperience	White et al. (2001)
Columbian ground squirrels	C	Juveniles < adults	Juveniles played more	Betts (1976)
Klipspringer	C	Juveniles < adults		Tilson (1980)
Hoary marmot	C	Juveniles > adults	Greater vulnerability to predation due to size and inexperience	Holmes (1984)
Spanish ibex	C	Juveniles < adults		Alados (1985)
Fallow deer	C	Juveniles = adults		Schaal and Ropartz (1985)
Yellow-bellied marmot	C	Juveniles > adults	Greater vulnerability to predation due to size and inexperience	Carey and Moore (1986)
Bighorn sheep	C	Juveniles < adults		Risenhoover and Bailey (1985)
Wedge-capped capuchin	C	Juveniles < adults		de Ruiter (1986)

TABLE 3.1 Effect of Age on the Amount of Time Allocated to Vigilance While Foraging in Birds and Mammals *(cont.)*

Species	Type of study*	Direction of effect	Probable causes	References
Eastern grey kangaroo	C	Juveniles < adults		Heathcote (1987)
Cape buffalo	C	Juveniles = adults		Prins and Iason (1989)
California ground squirrel	C	Juveniles < adults		Loughry and McDonough (1989)
Columbian ground squirrel	C	Juveniles < adults		MacHutchon and Harestad (1990)
Wedge-capped capuchin	C	Juveniles < adults		Fragaszy (1990)
Eastern grey kangaroo	C	Juveniles = adults		Colagross and Cockburn (1993)
African elephant	C	Juveniles = adults		Burger and Gochfeld (1994)
African buffalo	C	Juveniles = adults		Burger and Gochfeld (1994)
Burchell's zebra	C	Juveniles < adult males		Burger and Gochfeld (1994)
Wildebeest	C	Juveniles < adult males		Burger and Gochfeld (1994)
Defassa waterbuck	C	Juveniles < adult males		Burger and Gochfeld (1994)
Uganda kob	C	Juveniles < adults		Burger and Gochfeld (1994)
Impala	C	Inconclusive		Burger and Gochfeld (1994)
Springbok	C	Juveniles = adults		Bednekoff and Ritter (1994)

(Continued)

TABLE 3.1 Effect of Age on the Amount of Time Allocated to Vigilance While Foraging in Birds and Mammals *(cont.)*

Species	Type of study*	Direction of effect	Probable causes	References
Yellow-bellied marmot	C	Juveniles < adults	Higher feeding requirements in juveniles	Armitage and Corona (1994)
Vicuña	C	Juveniles < adult males		Vilá and Cassini (1994)
Nine-banded armadillo	C	Juveniles = adults	Adults have other targets of vigilance	McDonough and Loughry (1995)
Golden marmot	C	Juveniles > adults		Blumstein (1996)
Yellow-bellied marmot	C	Juveniles < adults	Inexperience	Li et al. (2011b)
California ground squirrel	C	Juveniles > adults	May learn from adults about predation threats	Hanson and Coss (2001)
Elk	C	Juveniles < adults	Higher feeding requirements in juveniles to survive winter	Childress and Lung (2003)
Bighorn sheep	C	Juveniles < adults		Mooring et al. (2004)
Dall sheep	C	Juveniles < adults	Inexperience and higher foraging needs	Loehr et al. (2005)
Elk	C	Juveniles < adults		Lung and Childress (2007)
Meerkat	C	Juveniles < adults	Lack of experience with predators	Hollen et al. (2008)
White-tailed deer	C	Juveniles < adults		Lark and Slade (2008)
Mountain goat	C	Juveniles = adults		Hamel and Côté (2008)
European rabbit	C	Juveniles < adults	Hungrier individuals were less vigilant; lack of experience and greater foraging needs	Monclús and Rödel (2009)

TABLE 3.1 Effect of Age on the Amount of Time Allocated to Vigilance While Foraging in Birds and Mammals *(cont.)*

Species	Type of study*	Direction of effect	Probable causes	References
Yellow-bellied marmot	C	Juveniles < adults	Higher feeding requirements in juveniles	Bednekoff and Blumstein (2009)
Cape ground squirrel	C	Juveniles = adults		Unck et al. (2009)
Spotted hyaena	C	Juveniles < adults	Lack of experience of juveniles	Pangle and Holekamp (2010a,b)
Mustached tamarin	C	Juvenile = adult		Stojan-Dolar and Heymann (2010a)
Yellow-bellied marmot	C	Juveniles > adults	In periods of heightened risk	Lea and Blumstein (2011a)
Nubian ibex	C	Juveniles > adults		Tadesse and Kotler (2011)
Ursine colobus monkey	C	Juveniles = adults		Teichroeb and Sicotte (2012)
Nubian ibex	C	Juveniles < adults		Tadesse and Kotler (2012)
Mouflon	C	Juveniles < adults	Traded-off vigilance for learning and energy acquisition	Benoist et al. (2013)
Bohor reedbuck	C	Juveniles < adults		Djagoun et al. (2013)
White-tailed deer	C	Juveniles < adults		Lashley et al. (2014)
Javan deer	C	Juveniles = adults		Pairah et al. (2014)
Mountain nyala	C	Juveniles > adults	Only in wet season	Tadesse and Kotler (2014)
Arctic ground squirrel	C	Juveniles > adults		Wheeler and Hik (2014)

*Longitudinal (L), cross-sectional (C), mixture of the two designs (L–C).

Two caveats are important to keep in mind when interpreting age-related changes in the expression of behaviour. First, changes with age might simply reflect maturation of the senses or the nervous system. As an example, in chub mackerel, a fish that forms large schools, individuals responded to cues of alarm from conspecifics a full 2 weeks after the onset of schooling behaviour, a finding thought to reflect the underdevelopment of the part of the brain involved in visual processing (Nakayama et al., 2007). Maturation is probably less of an issue in the above survey results since the juvenile category rarely included very young individuals.

Second, age-related changes might reflect adaptive changes to different selection pressures. Pre-programmed changes of this kind need not reflect an interaction with environmental stimuli (Lea and Blumstein, 2011b). Nestlings in many species of precocial birds crouch on the ground when threatened and rely on camouflage to escape predators (Tinbergen, 1953). As adults, the same individuals are no longer camouflaged and fly away upon alarm. Young individuals simply lack the ability to fly and must rely on other means to escape predators. Different selection pressures at different life stages can favour the evolution of different means to tackle the same problem.

Lower vigilance in juveniles than in adults has been ascribed to two mechanisms, namely, lack of experience in assessing or recognising predation threats and higher feeding requirements (Lea and Blumstein, 2011b). Lack of experience is no surprise given that juveniles are typically sheltered from predators by their parents. Higher feeding requirements in juveniles typically reflect their poorer foraging efficiency, an oft-cited cause of age-related changes in vigilance in the surveyed studies (Table 3.1). Juveniles also tend to have higher physiological needs due to their small size and growing requirements, which could cause a reduction in vigilance to accommodate the extra foraging time.

A particularly enlightening study on the development of vigilance featured the thirteen-lined ground squirrel, a rodent species in which juveniles devote less time to vigilance than adults (Arenz and Leger, 2000). To address the feeding requirements hypothesis, researchers provisioned juveniles with extra food. Supporting the idea that low vigilance reflects higher feeding requirements, juveniles supplemented with low-energy food showed less vigilance than those that received high-energy food (Fig. 3.7), a finding also documented in juvenile European rabbits (Monclús and Rödel, 2009). In a further study with the squirrels, which addressed this time the inexperience hypothesis, adults appeared more adept at judging risky situations than juveniles (Arenz and Leger, 1997a), suggesting that lack of experience with predators may also play a role in the lower vigilance of juveniles in this species.

The lack of experience of juveniles in recognising threats certainly fits the idea that the first step in vigilance is the ability to determine what to look for. Juveniles can form an association between a stimulus in the environment and a threat either through their own experience or by watching other group members. Once an association is made, individuals can use vigilance to detect such threats

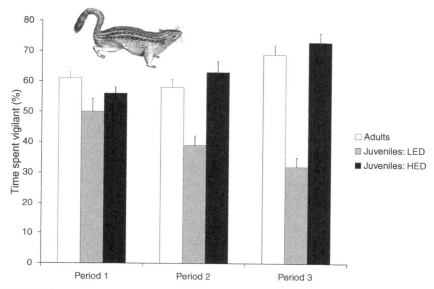

FIGURE 3.7   Feeding requirements and vigilance in juveniles. Juvenile thirteen-lined ground squirrels supplemented with a high-energy diet (HED) maintained the same level of vigilance as adults at three different time periods during the experiment, but were more vigilant than juveniles supplemented with a low-energy diet (LED). Mean and standard errors are shown. *(Adapted from Arenz and Leger (2000).)*

in the future. Supporting this idea, young meerkats in the African savannah produced fewer alarm calls than adults, but misjudged threats more often (Hollen et al., 2008). Similarly, young dwarf mongooses not exposed to the vigilance behaviour of adults failed to show appropriate vigilance behaviour later in life (Rasa, 1989). In Belding's ground squirrels, time spent vigilant by juveniles after an alarm call tended to mirror the pattern of reaction of their mothers (Mateo, 2014). This correlation might reflect learning from adults but also inherited response profiles.

Learning also appears to apply to social threats. In a longitudinal study, young rhesus monkeys showed appropriate responses to threatening facial patterns at 9 months of age but not 6 months earlier (Mandalaywala et al., 2014). Gradual acquisition of skills to recognize social threats also depended on the social experience of infants. Infants raised by higher-ranking and more protective mothers displayed greater vigilance for such threats than those from lower-ranking and less protective mothers, suggesting a role for parenting style.

By contrast to the inexperience hypothesis, one hypothesis predicts that juveniles should devote more, rather than less, time to vigilance than adults. Higher vigilance in young animals would reflect their greater vulnerability to predation caused by small size and limited escape ability (Holmes, 1984; Loughry, 1992). This hypothesis assumes that young animals are able to show adult levels of vigilance and recognise threats at an early age. One difficulty here is that parents could compensate for the greater vulnerability of their offspring by providing extra vigilance (Chapter 4). This could, in turn, bring vigilance down in juveniles, that is in the opposite direction to that predicted by the vulnerability hypothesis.

Many studies reported no effect of age on vigilance. Low statistical power might be the simplest explanation for many of these negative findings. However, I note that most studies focused on pre-emptive vigilance in typically low-risk situations. Greater vulnerability to predation might have more impact on vigilance during periods of heightened risk. If this is correct, age-related differences should be more pronounced in high-risk situations, explaining why age fails to influence vigilance when risk is lower. In line with this expectation, juvenile and adult yellow-bellied marmots showed the same level of vigilance during normal, low-risk foraging. Juveniles, however, responded more often to broadcasted alarm calls and also foraged less after the alarm calls than adults (Lea and Blumstein, 2011a). This study highlights the need to examine ontogenetic changes in vigilance in low- and high-risk situations.

Changes in vigilance with age need not be limited to the juvenile phase. Many species acquire foraging and reproduction skills beyond the juvenile stage, and the ability to respond to predators is often acquired over many years (Stankowich and Blumstein, 2005). Suggestive of changes in vigilance related to age in older individuals, first-time mothers in Belding's ground squirrels invested less in vigilance than multiparous females (Nunes, 2014). While it might be the case that young mothers have greater foraging needs, which would explain their lower vigilance, a case can also be made for learning. Lifelong adjustments in vigilance are worthy of more research.

3.4 EVOLUTIONARY ORIGIN OF ANIMAL VIGILANCE

The evolutionary origin of animal vigilance has received little attention and thus remains murky. This should come as no surprise given that animal vigilance leaves no trace in the fossil record. The only option left is to compare different species in the hope of retracing the different steps taken during evolution. With the caveat that some major steps in evolution may be missing from the available set of species, and that the species that we see today may have evolved from the presumed ancestral state that we seek to identify, it may be possible to retrace the development of a trait from its origin to present-day status by comparing different species.

Two avenues are possible to investigate the evolutionary origin of vigilance. The first focuses on anatomical structures underlying vigilance. Variation amongst species in particular anatomical features can be related to environmental factors associated with the need for vigilance. Two structures are particularly amenable for this type of analysis. Recall that the amygdala has emerged as a key player in coordinating the vigilant state, at least in primates. Variation amongst species in amygdala size could be related to need to maintain vigilance. This is not too far-fetched a concept as other structures in the brain like the hippocampus, which is associated with the formation of memory, vary in size in predictable fashion from species to species and even between the sexes in response to the need to process memory more efficiently (Biegler et al., 2001). The amygdala, or equivalent brain areas in different species, would hypothetically be more developed in species where predation and social risk matter most.

Another trait amenable to such an analysis is the eye. As more and more information on the visual fields of animals becomes available (Martin, 2007), specific features of the eyes could be related to the need for vigilance. The occurrence of horizontal streaks in the retina of avian species living in open habitats is a good example (Tyrrell et al., 2013). Horizontal streaks may have evolved in response to the need to monitor the horizon for predators. One would be able to retrace the steps in the evolution of vigilance by looking at how visual streaks evolved in species with different needs to monitor the horizon. Specializations in anatomical structures involved in the detection of predators are also apparent in acoustic vigilance. In moths that detect bat predators with their ears, species with ears more sensitive to the ultrasound frequencies used by bats to localize their prey are less susceptible to predation (Fullard, 2001). In all the above cases, the door is open to speculate about the ancestral state from which modern traits associated with vigilance evolved.

The second avenue to determine the evolutionary origin of animal vigilance focuses on the actual level of vigilance shown by different species rather than the structures that produce vigilance. Comparative analyses map the evolution of a trait on a phylogenetic tree, which displays presumed relationships amongst species over evolutionary times (Harvey and Pagel, 1991). This allows us not only to assess putative correlations with various ecological factors, but also to evaluate the ordering of events during evolution. A comparative analysis of this kind only makes sense when there is heritable variation in the investigated trait. The nightmare scenario in a comparative analysis happens if most of the variation in a trait reflects non-heritable sources like the environment.

Is there any indication that vigilance has a heritable component? As far as I know, only one study has directly investigated the heritability of vigilance behaviour (Blumstein et al., 2010). In yellow-bellied marmots, heritability of vigilance actually proved quite low. Much of the variation in vigilance was caused by extrinsic factors, and only 8% of the variation amongst individuals could be attributed to genes. Indirect evidence for heritability comes from recent studies that documented the occurrence of stable and distinct individual vigilance patterns in some species (Carter et al., 2009; Couchoux and Cresswell, 2012; Rieucau et al., 2010). This research will be covered in the following chapter.

I carried out a comparative analysis assuming a genetic component to some of the observed variation in vigilance amongst species (Beauchamp, 2010a). I focused on two characteristics of vigilance in birds readily available from the literature, namely, the proportion of time spent vigilant when alone, which

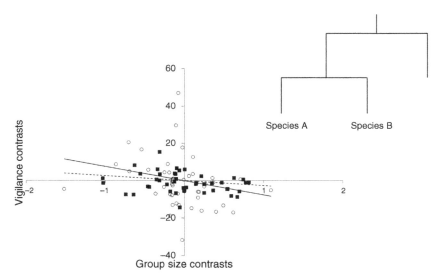

FIGURE 3.8   **A comparative analysis of vigilance.** In birds, an increase in group size over evolutionary times was associated with a decrease in asymptotic vigilance in vegetarian species (open circles and thick regression line) but not in carnivorous species (filled squares and dashed regression line). Asymptotic vigilance refers to the proportion of time spent vigilant in the largest groups of one species. Phylogenetic contrasts are shown for group size and vigilance. The inset shows a partial phylogeny with two species. Subtracting species values at each fork in the tree provided contrast values, which were standardized for comparisons across the whole phylogeny. *(Adapted from Beauchamp (2010a).)*

controls for inter-specific variation in group size, and the proportion of time spent vigilant in the largest groups, which I referred to as asymptotic vigilance because vigilance tends to level off as group size increases. I did not find any association between ecological factors and time spent vigilant when alone. However, asymptotic vigilance decreased with group size and increased with body mass in vegetarian but not in carnivorous species (Fig. 3.8), suggesting that avian species evolved a level of vigilance that reflects the risk they typically encounter. Whether evolution followed similar steps in mammalian species remains to be investigated.

3.5   CONCLUSIONS

Questions regarding the causation, development, and evolutionary origin of animal vigilance have all received little attention. The recent shift in animal behaviour research, from mostly functional analyses to a more encompassing framework including proximate considerations (Owens, 2006), bodes well for future research on these neglected issues.

Nevertheless, providing answers to these questions will be challenging for several reasons. Long-term studies are needed to establish the heritability of vigilance, but these studies are often difficult to fund (Clutton-Brock and Sheldon, 2010). Intensive sampling of the same individuals is required to perform longitudinal studies investigating the development of vigilance, but this is clearly difficult for many mobile species. Measurements of sensory ability, involving the eye or other senses, typically require expensive laboratory experiments in confined settings. As a consequence, few laboratories can perform such measurements, and for practical reasons these laboratories tend to focus on only a handful of amenable species.

How brain lesions influence vigilance has so far only been studied opportunistically in human subjects. Brain lesions can be performed on laboratory animals, but cannot be ethically done on wild animals whose survival would be jeopardized. Brain imaging studies, for practical reasons, are limited to humans and laboratory species that can fit in the machines. The scope for brain studies appears quite limited. Measurements of circulating hormones involved in vigilance can be done in the field, but experimental manipulation of hormone levels may be tricky ethically as increased mortality should be expected if anti-predator vigilance is affected. Our ability to meet the above challenges will dictate the pace of development in answering these non-functional questions.

Investigating vigilance in human subjects is a promising avenue for future research. One clear benefit is that we can actually ask the subjects to tell us about the target of their vigilance. Only a handful of studies on human vigilance have been carried out, mostly in semi-natural settings like cafeterias. There is a need to investigate vigilance in different settings perhaps involving indigenous populations in more natural environments. Laboratory investigations of vigilance in humans have a long history. However, laboratory research could benefit from using stimuli that are more clearly related to predation or social risk (Löw et al., 2008). The possibility of using brain-imaging techniques certainly adds to the appeal of human vigilance research.

Chapter 4

Drivers of Animal Vigilance

4.1 INTRODUCTION

The scientific naming of species tends to be quite conservative, and typically emphasizes striking external characteristics of a given species. Occasionally, the name of the researcher who first identified the species is used. A species may also be dedicated to a loved one or to benefactors. A species of lemur in Madagascar, the Bemaraha woolly lemur (*Avahi cleesei*) was named after the actor John Cleese of Monty Python's fame in recognition of his conservation efforts (Thalmann and Geissmann, 2005). A more regrettable dedication includes a species of blind cave beetle named after Hitler (Morrison, 2014). However, few species can boast of being named after a president of the United States of America. The Roosevelt elk (*Cervus canadensis roosevelti*), a large ungulate of North America's Pacific Northwest, was indeed named after the famous president Theodore Roosevelt who was behind the initiative to create a national park in the Mount Olympus area of Washington State where the elk is particularly abundant.

The Roosevelt elk is but one of several recognized sub-species of the elk, a species with a worldwide distribution. In western North America, the elk has been the subject of much research especially in the Yellowstone National Park where wolves, historically one of their main predators, were reintroduced in the winter of 1994–1995, helping recreate the landscape of fear that prevailed in earlier times (Laundré et al., 2001).

This new landscape of fear is actually quite fragmented as some areas are visited by wolves more often than others. Spatial heterogeneity in the abundance of wolves provides a golden opportunity to examine the effect of predation pressure on vigilance. Even in areas where wolves pose a threat, some parts are still safer than others. Grazing in meadows away from the wooded cover that provides protection against attackers is particularly risky for elk. Vigilance is thus expected to increase further away from protective cover. Elk are not totally defenceless and can alter predation risk to some extent by foraging in groups of different sizes. Larger groups provide more protection against predators, and would be expected to form in riskier areas. In large groups, elk may be able to relax vigilance to concentrate on foraging.

Elk also face threats from conspecifics. During the rutting season, males fiercely compete with one another for mating rights and also frequently harass

Animal Vigilance. http://dx.doi.org/10.1016/B978-0-12-801983-2.00004-8

81

females. Vigilance aimed at conspecifics may thus be a large component of over-all vigilance in both males and females, and could vary predictably from season to season and even between the sexes. Empirical research in the new landscape of fear has corroborated several of these predictions (Childress and Lung, 2003; Creel and Winnie, 2005; Laundré et al., 2001; Liley and Creel, 2008; Wolff and Van Horn, 2003).

Elk vigilance provides a good example of a behavioural pattern expected to vary with a host of ecological and social factors. I refer to any factor that has the potential to influence vigilance as a driver of vigilance. For elk, putative drivers of vigilance include group size, distance to protective cover, sex, and wolf density. Drivers of vigilance modulate the risk experienced by individuals. If vigilance favours early detection of threats, one would expect these drivers to influence vigilance levels in predictable fashion. In the following, I examine drivers of vigilance associated with social risk and with predation risk. In view of the vast amount of research on the effect of group size on vigilance, I devote the Chapters 5 and 6 to this factor and focus here on the remaining factors. I already covered the effect of age in Chapter 3, and will not consider it further here.

4.2 DRIVERS ASSOCIATED WITH SOCIAL RISK

Drivers of vigilance associated with social risk include factors that affect the nature and frequency of threats caused by conspecifics. Many of these factors play an important role during the reproductive season, but can also be influential at other times of the year.

4.2.1 Sex

Sexual differences in vigilance can arise through many factors only some of which are related to social risk. For instance, the sexes might differ in body size or physiological needs arising from reproduction; such factors influence vigilance primarily through differences in perceived predation risk or food require-ments. Here, I focus on threats caused by conspecifics, and how individuals of the two sexes can adapt vigilance accordingly.

Several mutually non-exclusive hypotheses have been proposed to explain sexual differences in vigilance in the context of social risk (Table 4.1). Not sur-prisingly, intra- and inter-sexual competitions take centre stage. Intra-sexual competition typically involves males competing with one another to maintain or gain access to females. Such competition may take the form of mate guarding or territory guarding. Higher vigilance by males has been reported in many species and is often ascribed to intra-sexual competition (Artiss and Martin, 1995; Baldellou and Henzi, 1992; Bertram, 1980; Burger and Gochfeld, 1994; Girard-Buttoz et al., 2014; Prins and Iason, 1989; Roberts, 1988; Rose and Fedigan, 1995; Tettamanti and Viblanc, 2014; Vilá and Cassini, 1994).

TABLE 4.1 Hypotheses Explaining Sexual Differences in Vigilance in Relation to Social Risk

Hypothesis	Explanation	References
Intra-sexual competition	Males are more vigilant to maintain or gain access to females	Burger and Gochfeld (1994); Jenkins (1944); Loughry (1993)
Inter-sexual competition	Females are more vigilant to detect and avoid prospective males	Cameron and Du Toit (2005); Manno (2007)
Male investment	Males are more vigilant to benefit their partners	Martin (1984); Wittenberger (1978)
Mate searching	Males are more vigilant to detect potential mates	Burger and Gochfeld (1988); Dunbar et al. (2002)

While intra-sexual competition appears a straightforward explanation for sexual differences in vigilance, it is important to examine other potential roles for vigilance. Consider the case of paired birds early in the reproductive season. Higher vigilance in paired males has been thought to decrease the risk of paternity loss as per the intra-sexual competition hypothesis. Using vigilance, males can detect and thus deter rival males from approaching their mates too closely. However, a conceivable alternative explanation is that males target their vigilance at predators rather than rivals. Recent studies of paired birds showed that correlates of predation risk actually explained more variation in vigilance than correlates of paternity loss, suggesting that higher vigilance in males serves to detect predators as well as competitors (Guillemain et al., 2003; Portugal and Guillemain, 2011) (Fig. 4.1).

Complicating matters, higher vigilance in paired males might also have evolved to provide benefits to their mates; this is known as the male investment hypothesis (Table 4.1). Higher vigilance by males would allow paired females to survive better and to accumulate more resources critical to reproduction. One difficulty here is whether higher vigilance in males evolved for this purpose or whether females incidentally take advantage of this high vigilance, which evolved for other purposes (such as deterring predators) (Squires et al., 2007). One way to assess whether males adjust vigilance to increase the success of their mates is to determine the degree of coordination of vigilance within the pair. By coordinating vigilance, a pair will increase the chances of detecting predators (see Chapter 7), which is a benefit to both partners, and also the amount of time allocated to feeding, which is particularly important for the female, who faces increased energy demands during reproduction. Coordination of vigilance was documented in pairs of white-tailed ptarmigan, providing support for the male investment hypothesis in this species (Artiss et al., 1999). A more direct way of testing this hypothesis would be to experimentally provide food to paired females and see whether males decrease their own vigilance when females can invest more time in vigilance.

FIGURE 4.1 **Vigilance in paired ducks.** In the Eurasian wigeon, paired males were more vigilant than unpaired males or females, and paired females. In this species, breeding males are gaudier than their more cryptic mates (inset). Means and standard error bars are shown. *(Adapted from Guillemain et al. (2003). Photo credit: Hirubo.)*

By contrast to the intra-sexual competition hypothesis, the inter-sexual competition hypothesis argues that females invest more time in vigilance to detect and thwart the sexual advances of males. In line with this prediction, female Utah prairie dogs in oestrous are particularly vigilant in detecting prospective males (Manno, 2007). In giraffes, males and females show the same level of vigilance overall, but a finer analysis revealed that females increased vigilance when their nearest neighbour was a larger male (Cameron and Du Toit, 2005). Prospective male giraffes, especially larger ones, are often aggressive and can cause foraging disruptions. By keeping an eye on threatening bulls, females can reduce the impact of such disturbances. In the alpine ibex, an increase in female vigilance in mating groups dominated by males also fits with this hypothesis (Tettamanti and Viblanc, 2014).

The occurrence of sexual differences in vigilance can certainly be indicative that at least one of the aforementioned three hypotheses operates. However, the absence of a sexual difference may indicate that social risk is, paradoxically, important but of similar magnitude in males and females. Returning to the giraffe example, large bulls represent a threat to both males and females, perhaps explaining the lack of an overall effect of sex on vigilance.

The various hypotheses for sexual differences in vigilance are not mutually exclusive and can act together to various degrees as was the case in the white-tailed ptarmigan. As another example, consider Thomas's langur, an Indonesian arboreal monkey species whose troops occupy overlapping territories. In overlapping areas, roaming males from one troop can challenge the territory owner from the other. During takeover attempts, roaming males often kill offspring of resident females. As predicted by the intra-sexual competition hypothesis, territorial males maintained more vigilance in overlapping areas (Steenbeek et al., 1999). Inter-sexual competition, however, also plays a role because females, especially those with infants at risk, tended to be more vigilant in overlapping areas. Male investment might also come into play because females tended to reduce their vigilance when close to the sentinel adult male. Overall, males maintained a higher level of vigilance than females, but the proper conclusion is not that females are not affected by social risk, but rather that males have more incentives to be vigilant.

The final hypothesis for sexual differences in vigilance involves mate searching, the use of vigilance to locate potential mates. In our own species, the finding that men in casual groups looked up more often than women was interpreted as a strategy to find potential friends or mates (Dunbar et al., 2002). Whether this mechanism applies to animals in the wild remains to be seen, but it could be relevant during periods of active mate search prior to the reproductive season.

The hypotheses reported in Table 4.1 typically predict more vigilance in males than females during the reproductive season. But in some species sexual differences persist even during the non-reproductive season (Reboreda and Fernandez, 1997), suggesting that monitoring competitors may not be the only reason why males are more vigilant. A further difficulty in the interpretation of sexual differences in social vigilance is that sex and dominance status are often intimately related. Individuals of one sex can displace those of the other from important resources or may be more aggressive in general. Since both sex and dominance can have an independent effect on vigilance (Waite, 1987b), it may prove difficult to disentangle their relative contributions. Another issue arises when the sexes are exposed to different levels of predation risk (Berger, 1991; Ekman and Askenmo, 1984), which could confound the predicted effect of sex as predation risk on its own can also affect vigilance (see Section 4.3).

4.2.2   Dominance

It has long been thought that vigilance betrays dominance relationships within a group (Chance, 1967; Jenkins, 1944). Dominance can influence vigilance the same way sex does: vulnerable individuals are expected to be more vigilant. Subordinates can benefit from vigilance by detecting or assessing threats posed by dominants; it allows them to escape more quickly or avoid these individuals or reduce their fear of attack. Dominance can influence vigilance in many contexts, but it has been mostly studied in relation to food competition. When competing for food, dominants can use aggression to steal food items from subordinates or displace them to other areas (Broom and Ruxton, 1998).

In line with the dominance hypothesis, many empirical studies have reported higher vigilance in subordinates (Blumstein et al., 2001a; Domènech and Senar, 1999; Dominguez, 2002; Ekman, 1987; Gaynor and Cords, 2012; Gende and Quinn, 2004; Haude et al., 1976; Keverne et al., 1978; McNelis and Boatright-Horowitz, 1998; Murton et al., 1966; Pannozzo et al., 2007; Pravosudov and Grubb, 1999; Ryan et al., 1996; Shepherd et al., 2006; Valone and Wheelbarger, 1998; Waite, 1987a,b).

Dominance often is correlated with other factors expected to influence vigilance independently, such as position in the group (Ekman et al., 1981; Hall and Fedigan, 1997; Hirsch, 2002; Murton et al., 1966), habitat (Brotons et al., 2000; Ekman, 1987; Hogstad, 1991; Krams et al., 2001; Matthysen, 1999), density of foragers (Hirsch, 2002), hunger (Krams, 1998), distance to cover (Hogstad, 1988b; Slotow and Coumi, 2000; Slotow and Rothstein, 1995b), age (Brotons et al., 2000) or sex (Gould et al., 1997; Pravosudov and Grubb, 1999; Waite, 1987b). All these factors can confound the relationship between vigilance and dominance, potentially masking or exacerbating the effect of dominance. Control for such confounding effects appears necessary to isolate the effect of dominance.

There are several exceptions to the aforementioned pattern, suggesting that the simple hypothesis of monitoring dominants is not always sufficient. Several studies thus revealed no effect of dominance on vigilance (Brotons et al., 2000; Cimprich and Grubb, 1994; Hall and Fedigan, 1997; Hamel and Côté, 2008; Hirsch, 2002; Krams, 2001; Liker and Barta, 2002; Slotow and Rothstein, 1995b; Trouilloud et al., 2004; Vahl et al., 2005; Wright et al., 2001a). In other cases, dominants actually maintained more vigilance (Gould, 1996; Gould et al., 1997; Hogstad, 1988b; Krams, 1998; Öst et al., 2007; Rose, 1994; Stahl et al., 2001; Treves et al., 2001). As an example, in one of the earliest empirical investigations of vigilance, large baboon males in a troop showed more vigilance than subordinates (Hall, 1960).

Why dominants would invest more in vigilance makes sense in particular contexts. Higher vigilance might be needed because dominants need to monitor competitors as well as predators (Rose, 1994). In family groups, for instance, higher vigilance by dominants can be used to protect kin from attacks by other families (Gould et al., 1997). Dominants might be more vigilant to defend their guarded resources against competitors (Mulder et al., 1995) or to monitor nearby subordinates to exploit their food discoveries (Hirsch, 2002; Stahl et al., 2001).

4.2.3 Distance to Neighbours

Results regarding the effect of sex and dominance on vigilance certainly suggest that social threats represent a major target of vigilance in many species of birds and mammals. Perhaps just as important as sex or dominance is the distance to competitors. Intuitively, dominants pose much less of a threat far away than nearby. Vigilance is thus expected to increase when threatening neighbours are closer. While most models of animal vigilance focus almost exclusively on total group size (see Chapter 5), distance to neighbours could potentially explain much more variation in vigilance than their numbers.

This point has been made most forcefully in primates. In many species of primates, the number of close neighbours explained more variation in vigilance than overall group size (Treves, 2000). However, a response to the density of neighbours need not indicate that vigilance is directed at threatening companions. Several mechanisms related to predation risk assessment also predict adjustments in vigilance when neighbours are closer (Section 4.3.4). What is unique about social risk is that vigilance is expected to increase when threatening neighbours are closer.

An increase in vigilance when neighbours are closer has been documented in some species, suggesting a role for social monitoring (Favreau et al., 2009; Hirsch, 2002; Kutsukake, 2006). For example, in troops of brown capuchin monkeys, vigilance increased when the number of companions within a 10-m radius increased (Hirsch, 2002) (Fig. 4.2). In this species, close neighbours increased the threat of attacks related to food.

Particularly suggestive of a role for social risk is the observation that the identity of close neighbours can also influence vigilance. This was the case in giraffes as we saw earlier: only the presence of a large bull nearby caused an increase in vigilance in both males and females (Cameron and Du Toit, 2005). Similarly, vigilance decreased in two primate species when the nearest neighbours were

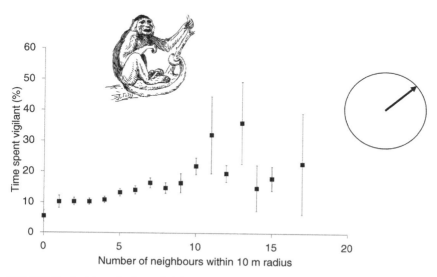

FIGURE 4.2   **Social monitoring and vigilance.** Vigilance in brown capuchin monkeys increased when the number of companions within a 10-m radius increased, suggesting more social monitoring when neighbours are closer. Means and standard error bars are shown. *(Adapted from Hirsch (2002).)*

familiar individuals or kin that posed fewer threats (Gaynor and Cords, 2012; Kutsukake, 2006).

4.3 DRIVERS ASSOCIATED WITH PREDATION RISK

4.3.1 Sex

Sex returns here as a determinant of vigilance, but this time in relation to perceived predation risk rather than social risk. The various hypotheses that have been put forward to explain sexual differences in this context typically emphasize differences in vulnerability to predation or in physiological needs, which cause one sex to take more risk or invest less in anti-predator vigilance (Table 4.2).

Sex-dependent vulnerability to predation can explain sexual differences in anti-predator vigilance. This could be the case if one sex is more conspicuous to predators (Guillemain et al., 2003; Hart and Freed, 2005; Lendrem, 1983b). In birds, for instance, predators may be disproportionately attracted to gaudier males (Huhta et al., 2003), explaining why males invest more in vigilance. However, the relationship between predation and conspicuousness is quite complex. Conspicuous traits in a prey animal could signal non-profitability to the predator instead (Götmark, 1993), in which case higher vigilance would be expected for the less, rather the more conspicuous sex. Many other traits can correlate with conspicuousness, such as a larger body size or differences in behaviour, all of which may tilt predation towards males independently of conspicuousness (Longland and Jenkins, 1987). Experimental manipulation of plumage brightness is needed to tease apart the relative contribution of all these factors in birds.

Sex-biased vulnerability to predation has also been linked to differences in body size in Eastern grey kangaroos. The small size of kangaroo females is thought to make them easier prey, explaining why females tended to be more vigilant than the larger males (Pays and Jarman, 2008). To be more convincing, this kind of explanation should rule out intra- and inter-sexual competition, which could also explain sexual differences in vigilance as explained earlier.

TABLE 4.2 Hypotheses Explaining Sexual Differences in Vigilance in Relation to Predation Risk

Hypothesis	Explanation	References
Vulnerability to predation	The more vigilant sex experiences a higher predation risk	Guillemain et al. (2003); Hart and Freed (2005); Lendrem (1983a); Pays and Jarman (2008)
Cost of reproduction	Reproduction demands force females to reduce vigilance	Lazarus and Inglis (1978); Rasa (1989)
Parental care	Females are more vigilant to protect vulnerable offspring	Burger and Gochfeld (1994); Caro (1987)

The ability to distinguish between anti-predator and social vigilance is also imperative because only anti-predator vigilance is expected to increase in the more vulnerable sex.

Any discussion of predation risk faces the challenge of assessing vulnerability. Predation risk is a highly abstract concept; it represents the hypothetical risk experienced by an individual investing no time or effort in avoiding predation (Lank and Ydenberg, 2003). Predation rate, one measurable consequence of predation risk, actually represents the net effect of predation after the deployment of anti-predator effort by prey animals. Therefore, the fact that gaudy males, say, are over-represented in the diet of a predator cannot be taken as evidence that males face a higher predation risk than females. Predators may actually prefer females, putting them at higher risk. However, females may be more difficult to catch after deploying their anti-predator tactics. Similarly, the fact that goat herders in Africa are rarely killed by lions does not mean that lions are not a threat; to the contrary, because lions are so dangerous, people are constantly on the lookout for them, and it is this high vigilance that reduces the risk of attack. Assessing predation risk for different classes of individuals, such as males and females when invoking the differential vulnerability hypothesis, is not easy. In the end, cafeteria-style experiments involving predators and prey dummies might be the only way to control for the deployment of anti-predator tactics in studies investigating predation risk.

Sex-dependent reproduction costs can also explain sexual differences in vigilance. This hypothesis proposes that individuals of the sex that invests the most in reproduction face greater nutritional challenges that reduce their ability to invest in anti-predator vigilance. Such a mechanism probably acts more forcefully if other individuals in the group compensate for the decrease in vigilance. This is the case in waterfowl where vigilance by males allows females to relax their own and feed more (Lazarus and Inglis, 1978). Similarly, in the sentinel system of dwarf mongooses, it is the relatively free males that tend to be more vigilant (Rasa, 1989).

A simple comparison between reproducing mothers and non-reproducing males has often been used to test the cost of reproduction hypothesis. Nevertheless, such a comparison can be difficult to interpret. This is because intra- and inter-sexual competition can also explain sexual differences in vigilance irrespective of reproduction costs. To address this issue, many studies compare instead reproducing and non-reproducing females to control for putative sexual differences in vigilance (Childress and Lung, 2003; Hamel and Côté, 2008; Manno, 2007; Rieucau and Martin, 2008; Ruckstuhl et al., 2003). Even this comparison is open to interpretation challenges; the two types of females might have access, say, to different types of resources or be drawn from pools of females with inherently different vigilance profiles. A clever way to address these issues would be to manipulate reproduction by allowing randomly selected females from the same group to reproduce or not during the same breeding season. In just this sort of experiment with Eastern grey kangaroos, reproducing

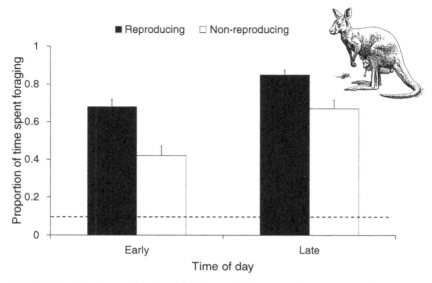

FIGURE 4.3 Cost of reproduction and vigilance. In Eastern grey kangaroos, reproducing mothers allocated more time to foraging than non-reproducing mothers early and late in the day, but with little consequences for anti-predator vigilance (dashed line). Experimental manipulation of reproduction status was performed with a fertility control agent. Means and standard errors bars are shown. *(Adapted from Cripps et al. (2011).)*

females surprisingly failed to reduce their vigilance despite the need to recoup substantial lactation costs (Cripps et al., 2011) (Fig. 4.3). While the generality of these findings has not been established, the simple expectation that reproduction costs can cause a decrease in vigilance may need to be revised.

Sex-dependent investment in parental care is the third and most widely cited mechanism at the origin of sexual differences in vigilance. This hypothesis proposes that the presence of vulnerable offspring forces individuals of the caring sex to allocate more time to vigilance. The parental care hypothesis also predicts a decrease in vigilance as offspring get older and less vulnerable. This prediction is examined in Section 4.3.2.

In many species of birds and mammals, higher vigilance in females, the caring sex, has thus been ascribed to the protection of offspring against predators (Benoist et al., 2013; Berger and Cunningham, 1995; Childress and Lung, 2003; Colagross and Cockburn, 1993; Lashley et al., 2014; Li et al., 2012; Li et al., 2009; Rieucau et al., 2012; Shi et al., 2011; Tadesse and Kotler, 2011; Wahungu et al., 2001; Xia et al., 2011). Many studies, however, reported no effect of parenting on vigilance (Ebensperger and Hurtado, 2005b; Gould, 1996; Hamel and Côté, 2008; Manno, 2007; Unck et al., 2009). Facing nutritional challenges, as per the cost of reproduction hypothesis, mothers might not be able to increase vigilance when caring for vulnerable offspring, which could explain the lack of effect of parenting on vigilance in the aforementioned studies.

4.3.2   Age and Number of Offspring

The presence of vulnerable offspring can certainly influence vigilance in birds and mammals. What happens to parental vigilance as offspring get older or in larger families? Predation risk probably decreases when offspring become more mobile and experienced with age (Chapter 3). By contrast, predation risk probably increases with the number of vulnerable dependents. The simple expectation would be that vigilance should decrease as offspring get older and be higher in larger families.

A decrease in vigilance as offspring get older has been documented in many species of birds and mammals (Caro, 1987; Laurenson, 1994; Loonen et al., 1999; Pangle and Holekamp, 2010b; Seddon and Nudds, 1994; Sullivan, 1988; Treves et al., 2003). In cheetahs, for instance, mothers became less aggressive and less vigilant when accompanied by older, less vulnerable cubs (Fig. 4.4). In general, the ability of offspring to detect and escape predators should increase with age. Less parental investment in vigilance as offspring get older parallels the decrease in vulnerability.

This pattern of decrease in parental vigilance with offspring age is not universal. Exceptions include black-tailed prairie dogs in which mothers actually became more vigilant with older offspring (Loughry, 1993). This is surprising because in this species younger offspring are in fact more vulnerable to predation. This unexpected pattern of vigilance is probably related to the energetic costs of lactation. High lactation costs in the early days of pup rearing probably

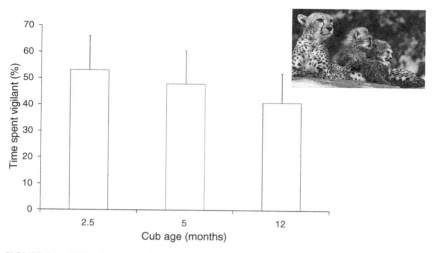

FIGURE 4.4   Offspring age and vigilance. The mean percentage of time spent vigilant by cheetah mothers decreased as their cubs aged. The range of cub ages for each bar was 1.5–3.5 months, 4.5–6.5 months, and 8–18 months, respectively. Means and standard deviation bars are shown. *(Adapted from Caro (1987). Photo credit: Tambako.)*

force females to devote most of their time to feeding at the expense of vigilance. As energy demands ease when offspring get older, females can devote more time to vigilance. In this species, offspring age is totally entwined with parental state, making it hard to disentangle their relative contributions to vigilance. Belding's ground squirrels also showed a similar increase in vigilance with offspring age (Nunes et al., 2000). But this increase in vigilance prevailed in both provisioned and non-provisioned females, suggesting that reduced physiological demands on females cannot fully account for increased vigilance in this species.

An increase in vigilance as offspring get older might actually be adaptive rather than a spurious consequence of parental state. The argument involves the value of offspring to parents from an evolutionary point of view. The probability that offspring reach reproductive age increases with age. As a result, older offspring are that much more valuable to parents aiming to pass their genes into future generations. Increased offspring value should favour greater parental investment (Trivers, 1972), which could take the form of increased parental vigilance. The parental investment hypothesis was invoked to explain the increase in parental vigilance as common eider ducklings aged (Öst et al., 2002). I note, however, that female eiders with older ducklings might also have more time to allocate to vigilance after recouping prior energetic losses. This possibility was not addressed.

Offspring vulnerability and offspring value predict opposite patterns of parental vigilance with offspring age. Parental state can also change according to offspring age, complicating matters even more. It is not so simple, after all, to predict how vigilance should change with offspring age.

I now turn to the effect of offspring number on parental vigilance. This relationship has mostly been investigated in birds because family size is typically small in mammals. In birds, several studies showed an increase in parental vigilance in larger families (Black and Owen, 1989; Forslund, 1993; Loonen et al., 1999; Sedinger et al., 1995; Sedinger and Raveling, 1990; Tinkler et al., 2007; Walters, 1982; Williams et al., 1994), but the pattern is not universal (Badzinski, 2005; Caithamer et al., 1996; Ge et al., 2011; Jónsson and Afton, 2009; Lazarus and Inglis, 1978; Lessells, 1987; Li et al., 2013; Ruusila and Poysa, 1998; Seddon and Nudds, 1994).

Two hypotheses can explain an increase in parental vigilance in larger families. First, larger families might inherently be more at risk. Large families could attract more predators, for example. The need to monitor rivals competing for resources might also increase in larger families (Tinkler et al., 2007). In the South American vicuña, a rodent species in which one male defends a territory with several females and their offspring, vigilance increased with family size because larger territories with more offspring attracted more rivals (Vilá and Roig, 1992). The second hypothesis proposes that larger families have a greater future reproductive potential, which makes them more valuable to parents (Trivers, 1972).

By contrast to the effect of offspring age on vigilance, family risk and family value both predict the same pattern of increase in vigilance with offspring number. The lack of relationship between offspring number and vigilance noted previously is that much more surprising. One possibility to explain the lack of effect of family size is simply that the umbrella of parental care can effectively cover more than one offspring at the same time. Adding more offspring thus requires no additional investment in vigilance (Lazarus and Inglis, 1986). Predator behaviour is a further consideration for this explanation. When predators can capture at most one offspring from the family during an attack, vigilance is not expected to increase with the number of offspring. By contrast, the risk of losing the whole family to a predator during one attack is expected to favour more vigilance in larger families. Differences in predator behaviour and vulnerability of offspring could explain some of the variation amongst species in the effect of offspring number on vigilance.

Parents with more offspring might naturally be better at foraging and at keeping an eye on predators at the same time. In this case, an increase in vigilance with family size could simply reflect parental quality. Manipulation of family size is necessary to break down this correlation. In the barnacle goose, where such an experiment was performed, vigilance still increased with family size (Loonen et al., 1999), highlighting the need to maintain more vigilance in families with more vulnerable or more valuable offspring.

4.3.3   Body Mass

Adult body mass can influence vigilance because predation risk may be size-dependent. All else being equal, a larger, better-defended prey animal should experience a reduction in predation risk. Earlier, we saw that this mechanism was invoked to explain sexual differences in vigilance in kangaroos. Lower vulnerability to predation might also be the reason why females in larger species of mammalian herbivores invest less in vigilance (Berger and Cunningham, 1988). However, many factors other than body mass can vary amongst species, including the type of forage and other life-history considerations, which might explain differences in vigilance across species.

Documenting vigilance as a function of body mass within the same sex and in the same species would provide a stronger test of the body mass hypothesis. A recent study was able to do just that by comparing vigilance in males of the alpine ibex, an ungulate species in which older males can weigh up to twice as much as younger ones. As predicted, vigilance decreased in heavier males, controlling for many potential confounding factors such as group size and forage quality (Brivio et al., 2014). Crucial evidence that heavier males were less likely to be preyed upon was unfortunately missing. Another consideration is that younger males, with a higher residual reproductive value, might be less willing to sacrifice safety (by decreasing vigilance) than older males with fewer reproductive years ahead of them.

4.3.4   Distance to Neighbours

Distance to neighbours returns as a driver of vigilance, but this time in relation to predation risk. Several mechanisms related to predation risk are spatially sensitive and thus relevant to vigilance. One hypothesis proposes that in a tighter group, the domain of danger of many individuals, that is, the area surrounding one individual in which this individual is more at risk of predation than the neighbours, will be reduced (Hamilton, 1971). Reduction in relative predation risk when neighbours are closer should favour a decrease in anti-predator investment like vigilance. Collective detection of threats and the dilution of risk that arises from the presence of alternative targets for a predator in a group, will also be more effective when individuals are closer, again causing a decrease in vigilance (Bednekoff and Lima, 1998b). I review these mechanisms more fully in Chapter 5. As far as predation risk is concerned, an increase in inter-individual distances is predicted to increase vigilance.

Consistent with this prediction, vigilance increased in many species when neighbours were further apart (Couchoux and Cresswell, 2012; Devereux et al., 2006; Di Blanco and Hirsch, 2006; Fernández-Juricic and Beauchamp, 2008; Fernández-Juricic et al., 2007a; Fernández-Juricic et al., 2004c; Fernández-Juricic et al., 2005; Gaynor and Cords, 2012; Marino and Baldi, 2008; Pöysä, 1994; Radford and Ridley, 2007; Randler, 2005a; Rolando et al., 2001; Smith et al., 2005; Smith et al., 2004; Smith and Cain, 2009; Treves, 1998; Treves et al., 2001; Unck et al., 2009; Valcarcel and Fernández-Juricic, 2009). In other species, vigilance showed no statistically significant relationship with inter-individual distances (Ciuti et al., 2012; Dominguez, 2002; Fairbanks and Dobson, 2007; Pays et al., 2009a; Randler, 2005c; Shi et al., 2011).

Distance to neighbours can be assessed visually by prey animals. However, a study in pied babblers suggested that assessment of individual spacing can be done just as well with vocal cues made during foraging (Radford and Ridley, 2007). Auditory assessment of density might be useful for species foraging in visually-cluttered habitats.

When a group is limited to a fixed area, any increase in group size should lead to a decrease in inter-individual distances. In this case, it is not clear whether vigilance decreases because neighbours are closer or because the group is larger. Group size and inter-individual distances need not be correlated when a larger group can spread out, but the potentially confounding effect of group size must be considered when evaluating the effect of inter-individual distances on vigilance. The most insightful studies in that regard maintain group size constant and only vary the distance between neighbours. Studies of this kind revealed a strong decrease in vigilance when neighbours are closer, consistent with the predation risk hypothesis (Fernández-Juricic and Beauchamp, 2008; Fernández-Juricic et al., 2007a; Fernández-Juricic et al., 2004c) (Fig. 4.5).

Close neighbours can reduce predation risk. But foraging too close to companions could be problematic. In particular, the level of food competition could

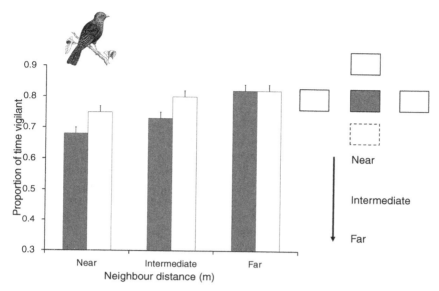

FIGURE 4.5 Distance to neighbours and vigilance. Vigilance in brown-headed cowbirds increased when neighbours were more distant. The trend was apparent for central (grey bars) and peripheral foragers (white bars). To increase neighbour distances, one cage at the edge of the group (dashed box) was moved at different distances from the centre of the group. Means and standard error bars are shown. *(Adapted from Fernández-Juricic and Beauchamp (2008).)*

rise if neighbours compete for limited resources. Because food competition can cause a decrease in vigilance (Clark and Mangel, 1986), low vigilance when neighbours are closer might conceivably be the outcome of increased food competition rather than reduced predation risk. A proper control for food competition is needed to ascertain the role of inter-individual distances on vigilance.

4.3.5   Visual Obstructions

Obstructions in the habitat can prevent individuals from acquiring information about predation risk. Vigilance in obstructed habitats is expected to increase to compensate for degraded information. In the case of acoustic vigilance, say, a rise in ambient noise levels might detract from the ability to acquire auditory information about approaching predators. In chaffinches, a small European passerine bird species, visual vigilance increased when the level of ambient noise was artificially raised (Quinn et al., 2006), suggesting that birds tried to compensate for reduced detection ability. A similar finding was documented in foraging mule deer exposed to anthropogenic noise (Lynch et al., 2015). Studies on the effect of obstructions on vigilance have mostly focused on visual rather than auditory vigilance. Here, I deal with visual obstructions, but the conclusions should generally apply to any type of vigilance.

Many habitat features can obstruct the field of view. Examples include stubble in a field (Murton and Isaacson, 1962), rocks on the shore (Metcalfe, 1984), other group members (Burger and Gochfeld, 1991b; Kluever et al., 2008), or dense vegetation cover (Ebensperger and Hurtado, 2005a). In one of the earliest empirical studies on vigilance in birds, Murton and Isaacson (1962) found that wood pigeons adopted the raised posture associated with vigilance more often in stubble fields, as would be expected if birds could not see their surrounding as well amidst the vegetation. In shorebirds, rocks on the beach can obstruct the field of view and prevent foragers from spotting flying predators. On rockier shores, shorebirds thus increased their vigilance either by scanning more frequently or by scanning longer (Metcalfe, 1984). This response is interesting because rocks might actually decrease predation risk by hiding birds from their predators. Obviously, these shorebirds did not consider the rocky shore safer. By contrast, fat sand rats perceived visually obstructive cover as safer and reduced their vigilance under cover (Tchabovsky et al., 2001). Such species differences probably reflect differences in the nature of the cover and the protection it provides to prey animals (Hall et al., 2013; Hannon et al., 2006).

As was the case in wood pigeons and shorebirds, many other studies in birds and mammals documented an increase in vigilance in more obstructed habitats (Blumstein et al., 2003; Devereux et al., 2006; Ebensperger and Hurtado, 2005a; Goldsmith, 1990; Guillemain et al., 2001; Javurkova et al., 2011; Lima, 1991, 1992; Underwood, 1982; Wheeler and Hik, 2014). Artificially created visual obstructions have also been useful to investigate the effect of obstructions on vigilance. By contrast to natural obstructions, artificially created visual barriers, such as small walls, can be presented under standardized conditions controlling for correlated extraneous factors. Echoing the aforementioned results with natural obstructions, all experimental studies with artificially-created visual barriers documented an increase in vigilance (Arenz and Leger, 1997a,b; Arenz and Leger, 1999; Bednekoff and Blumstein, 2009; Butler et al., 2005; Fernández-Juricic et al., 2005; Iribarren and Kotler, 2012; Lima and Bednekoff, 1999a; Makowska and Kramer, 2007; Quirici et al., 2008; Whittingham et al., 2004) with one exception (Foster-McDonald et al., 2006).

A recent and particularly insightful study investigated the responses of yellow-bellied marmots to both naturally occurring and artificially created visual barriers (Bednekoff and Blumstein, 2009). Naturally dense vegetation cover can block sightlines in species like marmots foraging low on the ground. Accordingly, marmots increased their vigilance in more cluttered meadows. In an apparatus that blocked the view from three sides, marmots compensated for peripheral visual obstruction by increasing vigilance when outside the apparatus. This combination of observational and experimental approaches increases our confidence that vigilance tactics are adapted to the level of visual obstruction.

In another study, however, visual barriers produced the opposite effect. House sparrows feeding on a platform on the ground actually reduced their vigilance when a small wall obstructed their field of view (Lima, 1987a). One possibility to explain this unexpected finding is that the wall actually provided cover against predators. Alternatively, the wall, which was relatively high for the birds, might have made it rather difficult to initiate a scan, resulting in a decrease in vigilance. The protective cover hypothesis might also explain other cases in which vigilance decreased with visual obstructions (Cowlishaw, 1994; Griesser and Nystrand, 2009; Hall et al., 2013; Hannon et al., 2006; Nersesian et al., 2012).

If areas distant from cover are more open, vigilance might be expected to decrease further away from cover. Early empirical work with house sparrows showed that contrary to expectations, individuals were actually more vigilant away from cover where intuitively their field of view should have been less obstructed (Barnard, 1980; Caraco et al., 1980b). However, an increase in vigilance away from cover makes sense for species that use cover as a refuge from predators. I return to this point in Section 4.3.6.

Visual obstructions can make it more difficult to detect approaching predators. However, obstructions could also reduce the ability to detect companions or competitors. Murton and Isaacson (1962), in fact, related the higher vigilance of wood pigeons in stubble fields to the difficulty of seeing one another rather than approaching predators, a point echoed subsequently by other researchers (Duncan and Jenkins, 1998; Elgar et al., 1984; Harkin et al., 2000). In a visually cluttered habitat, fewer group members are visible at any given time and foragers appear further apart. This skewed perception of group size and density alone could lead to an increase in vigilance. In addition, visual obstructions could make it more difficult to monitor other group members for alarm signals, again causing more vigilance (Harkin et al., 2000). More work is needed to tease apart the contribution of all these different factors to the increase in vigilance in cluttered habitats.

Species can adapt to clutter by altering their vigilance tactics. Some species become more vigilant and others can even modify their vigilance postures. For instance, terrestrial quadrupedal species foraging in dense vegetation resort to bipedal vigilance to get a better field of view (Bednekoff and Blumstein, 2009; Ebensperger and Hurtado, 2005a). Blocked sightlines, however, might be a curse for less adaptable species. If vigilance of any kind is ineffective in a cluttered habitat, individuals may simply decrease vigilance altogether and feed faster (Embar et al., 2011; Scheel, 1993).

4.3.6 Distance to Cover

Cover can be obstructive or protective, and the distinction between the two is essential to predict how distance to cover influences vigilance (Lazarus and Symonds, 1992). In the shorebird species that I study, the semi palmated sandpiper, individuals forage on exposed mudflats bordered by thick forest cover. Falcons, one of their main predators, launch attacks from the tree cover to conceal their fast approach. Upon attack, sandpipers have nowhere to hide on the exposed mudflats and form tight airborne defensive clusters to increase safety (Beauchamp, 2010c; Dekker et al., 2011). Tree cover in sandpipers is obstructive. Further away from obstructive cover, less vigilance is needed because animals can detect approaching predators more quickly giving them a head start when escaping.

House sparrows, as we have seen earlier, fly to nearby bushes when escaping from a predator. Dall's sheep can outrun terrestrial predators by seeking terrain with steep slopes. Bushes and steep slopes are two examples of protective cover. When protective cover is further away, more vigilance allows prey animals to detect predators sooner giving them more time to escape.

Cover should not be classified as protective or obstructive based on the observation that animals are more or less vigilant away from it. To avoid a circular argument, it is better to determine the visual properties of cover *a priori* and to learn about the usual escape tactics of prey animals. On the basis of this information, cover can be characterized as protective or obstructive, and vigilance can then be documented after the fact.

Many empirical studies examined the effect of distance to cover on vigilance in birds and mammals. For protective cover, an increase in vigilance with distance to cover has been the overwhelming pattern documented in the literature (Barnard, 1980; Barta et al., 2004; Blumstein and Daniel, 2002; Blumstein et al., 2001b; Caraco et al., 1980b; Carrascal et al., 2001; Cassini, 1991; Ciuti et al., 2012; Colagross and Cockburn, 1993; Coolen and Giraldeau, 2003; Diaz and Asensio, 1991; Edwards et al., 2013; Ekman, 1987; Frid, 1997; Hogstad, 1988b; Iribarren and Kotler, 2012; Lazarus and Symonds, 1992; Liley and Creel, 2008; Lima, 1987a; Matson et al., 2005; Pascual and Senar, 2013; Tadesse and Kotler, 2014; White and Berger, 2001; White et al., 2001; Wikenros et al., 2014). Frid's study on Dall's mountain sheep nicely illustrates how vigilance increases further away from steep slopes (Fig. 4.6).

There are quite a few exceptions to this pattern (Barnard and Stephens, 1983; Blumstein et al., 2003; Creel et al., 2014; Elgar, 1986a; Fanson et al., 2011; Favreau et al., 2009; Lima, 1988; McDonald-Madden et al., 2000; Mella et al., 2014a; Pascual et al., 2014; Pays et al., 2009a; Pöysä, 1994; Randler, 2003; Slotow and Coumi, 2000; Slotow and Rothstein, 1995b; Wahungu et al., 2001). Confounding effects caused by uncontrolled factors can be a reason for the negative findings (more on this will be discussed later). Another possibility is that to reduce risk prey animals prefer to decrease overall time spent away from cover, by decreasing vigilance, rather than increase the proportion of time spent vigilant (Beauchamp and Ruxton, 2007; Mella et al., 2014a).

There have been fewer studies with obstructive cover. Most studies documented a decrease in vigilance further away from obstructive cover (Beauchamp, 2014a; Fuller et al., 2013; Halofsky and Ripple, 2008; LaGory, 1986; Schutz and Schulze, 2011), although there are some exceptions (Couchoux and Cresswell, 2012; Sansom et al., 2008).

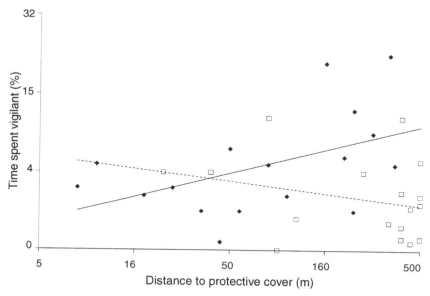

FIGURE 4.6   Vigilance and distance to protective cover. The proportion of time spent vigilant (in log scale) increased in Dall's sheep foraging further away from the protection of cliffs. The pattern of increase was more apparent in small groups of 2–4 (filled diamonds) than in larger groups of 30–100 (open squares). Trend lines are shown for each type of groups. *(Adapted from Frid (1997).)*

Distance to cover is often correlated with uncontrolled factors related to vigilance, making it difficult to interpret the results of observational studies. In many bird species, subordinates are often forced to forager further away from cover (Hogstad, 1988b; Pascual et al., 2014; Schneider, 1984; Slotow and Rothstein, 1995b), which causes a correlation between distance to cover and dominance status. As described earlier, subordinates tend to be more vigilant than dominants, perhaps explaining why vigilance increases further away from cover. Another potential confounding factor is group size. As individuals focus more on detecting predators away from protective cover, fighting amongst individuals might be expected to decrease, perhaps allowing an increase in group size (Caraco et al., 1980b). For this or other reasons, group size and distance to cover are often positively correlated (Barnard and Stephens, 1983; Creel and Winnie, 2005). An increase in group size further away from protective cover, and the concomitant increase in safety, might hide the expected increase in vigilance (Wahungu et al., 2001). In other cases, group size or inter-individual distances decreased with distance to obstructive cover (Beauchamp, 2010c; Fuller et al., 2013; Whitfield, 2003), again making it difficult to isolate the effect of distance to cover.

Experimental control of such confounding factors is needed to provide less ambiguous results. This has only been done in a handful of studies. In ingenious experiments, researchers made artificial bushes with pieces of wood and located cover where needed (Slotow and Rothstein, 1995a) or changed the properties of cover to make it protective or obstructive (Lazarus and Symonds, 1992).

Predictions about the relationship between distance to cover and vigilance rely on the assumption that prey animals are primarily exposed to one source of predation. With one type of predator, it is fairly easy to determine whether cover is obstructive or protective. But what happens when prey animals face different predators, each with their own attack tactics? In this case, it might not be obvious to predict how distance to cover influences vigilance. Vegetation cover for agile wallabies in Australia is typically viewed as protective; eagles tend to attack from the open and wallabies retreat to dense vegetation for safety (Blumstein et al., 2003). However, in some areas, crocodiles also lurk close to vegetation cover, making cover obstructive this time. In areas with both types of predators, cover is not necessarily protective or obstructive. A detailed knowledge of all threats faced by prey animals is needed to make accurate predictions regarding the effect of distance to cover on vigilance.

4.3.7 Position in the Group

Solitary foragers can only reduce predation risk by increasing vigilance or by selecting a safer niche. Group foragers, however, have the additional option of relying on one another to detect predators. Despite greater safety, not all individuals in a group are expected to face the same predation risk. The spatial position of a forager in a group has emerged as one of the most important drivers of vigilance in animals. In groups that are large enough, individuals can occur at the edges or in more central positions. At the edges, individuals have neighbours on only one side while those in the middle have neighbours on all sides. There are more complex ways to define spatial position (Stankowich, 2003), but this simple definition suffices here.

Vigilance is expected to vary with spatial position for a variety of reasons. Three non-mutually exclusive hypotheses are discussed here. The first hypothesis involves social vigilance and the need to monitor competitors while the other two deal with anti-predator vigilance. The conspecific monitoring hypothesis applies to species living in groups spatially segregated along dominance lines. Long ago, Murton and his colleagues noticed that wood pigeons at the edges of a group maintained more vigilance than those behind (Murton et al., 1966). Higher vigilance was attributed to the need to monitor dominant birds located inside the group.

The domain of danger hypothesis proposes that certain positions in the group offer more protection against predators. Hamilton first defined the domain of danger as the exclusive area surrounding an individual in which this individual, but no one else, is more likely to be attacked (Hamilton, 1971). If a predator attacks from the outside of the group, individuals at the edges are expected to have a larger domain of danger because they have no neighbours on one side to buffer them from predators (Vine, 1971). Greater exposure to predation threats at the edges of the group would select for increased vigilance.

The collective detection hypothesis proposes that individuals at the edges benefit less from the alarm responses of other individuals because they tend to be further apart from one another (Lazarus, 1978). Impaired collective detection would select for increased vigilance. As a mechanism to explain spatial variation in vigilance, impaired collective detection has received less attention that the domain of danger argument. Nevertheless, theoretical work showed that this mechanism alone could explain why vigilance is higher at the edges of a group (Proctor et al., 2006). In addition, because edge individuals tend to be closer to an attacking predator, an increase in vigilance is likely to be more effective in detecting attacks than relying on collective detection (van der Post et al., 2013).

Many empirical studies in birds and mammals showed that foragers at the edges of groups are indeed more vigilant (Alados, 1985; Beauchamp, 2014a; Beauchamp and Ruxton, 2008; Black et al., 1992; Blanchard et al., 2008; Carbone et al., 2003; Colagross and Cockburn, 1993; Couchoux and Cresswell, 2012; Di Blanco and Hirsch, 2006; Dias, 2006; Djagoun et al., 2013; Ekman et al., 1981; Fernández-Juricic and Beauchamp, 2008; Inger et al., 2006; Inglis and Lazarus, 1981; Jennings and Evans, 1980; Keys and Dugatkin, 1990; Klose et al., 2009; Kuang et al., 2014; Lazarus, 1978; Lipetz and Bekoff, 1982; Owen, 1972; Prins, 1996; Radford and Ridley, 2007; Shi et al., 2011; White-Robinson, 1982). Spatial position also influences vigilance when animals are sleeping (Dominguez, 2003; Lung and Childress, 2007; Rattenborg et al., 1999a) or preening (Dominguez and Vidal, 2007; Randler, 2005c). Some studies reported no effect of spatial position on vigilance (Ciuti et al., 2012; Dalmau et al., 2010; Hall and Fedigan, 1997; Hirsch, 2002; Liley and Creel, 2008; Randler, 2003; Sansom et al., 2008; Tkaczynski et al., 2014).

The aforementioned studies typically invoked the domain of danger hypothesis to explain the effect of spatial position on vigilance. However, the crucial evidence showing that attack rate or survival varies with spatial position is often if not always missing (Stankowich, 2003). Few have considered the relevance of the other two alternative hypotheses. A recent study in fact provided support for the conspecific monitoring hypothesis. In common eiders, vigilance was actually higher in the middle of the brood (Öst et al., 2007). Higher vigilance might be needed to monitor peripheral females that often interfered with brood-rearing activities.

Spatial variation in vigilance has interesting consequences for species living in groups. In larger groups, the number of individuals at the edges represents a smaller proportion of the total number of foragers (i.e. the periphery-to-area ratio decreases). Therefore, as group size increases, a randomly selected individual is less likely to be drawn from the edges where vigilance is higher. If vigilance at the group level is measured from a set of randomly selected individuals, vigilance will tend to spuriously decrease with group size because proportionately fewer more vigilant individuals are sampled from larger groups. Such

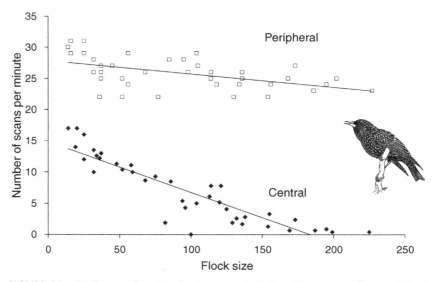

FIGURE 4.7 **Vigilance and position in the group.** In flocks of European starlings, peripheral individuals (open squares) maintained a higher vigilance than central (filled diamonds) companions, but both types of individuals reduced their vigilance in larger groups. Trend lines are shown for data at each spatial position. *(Adapted from Jennings and Evans (1980).)*

a decrease in vigilance with group size can happen even if each individual actually maintains the same vigilance in groups of different sizes (Lazarus, 1978; Lipetz and Bekoff, 1982).

To document changes in vigilance with group size, it is important to determine whether individuals at the edges and at the centre of the group alter their vigilance in groups of different sizes. In European starlings, vigilance at the group level decreased with group size (Jennings and Evans, 1980). Ruling out the aforementioned geometric argument, vigilance decreased with group size both at the edges and at the centre of the flocks (Fig. 4.7).

Spatial position is often correlated with extraneous factors. Unwanted correlations of this kind can confound the relationship between spatial position and vigilance. Empirical studies implicitly assume that different types of individuals are spread evenly across the group. In the wood pigeon study mentioned earlier, this assumption was not met because subordinates tended to occupy peripheral positions, which on its own might explain higher vigilance at the edges. Dominance status and spatial position in the group are also entwined in other species (Black et al., 1992; Brotons et al., 2000; Ekman et al., 1981; Whitfield, 1985).

Other factors are also correlated with spatial position in the group. For instance, in groups composed of more than one species, some species preferentially occupy peripheral positions (Bshary and Noë, 1997; Caldwell, 1986), again violating the assumption of even spread. In single-species groups, some age-sex classes of individuals were also more common at the edges (Brotons et al., 2000; Fragaszy, 1990; Hall and Fedigan, 1997). In some species of fish, the parasite burden of peripheral individuals tends to be higher than those in central positions, which might influence the allocation of time to different activities including vigilance (Barber et al., 2000). In one species of monkeys, individuals with better colour vision, which probably helps to detect predators, also occurred more peripherally (Smith et al., 2005). All the aforementioned differences can confound or modify the effect of spatial position on vigilance.

Peripheral individuals often tend to be further apart, as suggested by the impaired collective detection argument described earlier. This factor alone could lead to an increase in vigilance, irrespective of spatial position in the group. When controlling for the spacing of individuals, vigilance in brown capuchin monkeys actually showed no relationship with spatial position (Hirsch, 2002). In terms of social vigilance, more crowding in the middle of the group could lead to more interactions and more monitoring (Dalmau et al., 2010). In view of all these potential effects on vigilance, the spacing of individuals should be taken into account when investigating the effect of spatial position on vigilance.

Another common confounding factor is food availability or more generally food competition (Barnard and Thompson, 1985; Carbone et al., 2003; Krause, 1994). In wood pigeons, for instance, the supply of fresh leaves decreased at the edges of the flocks (Murton et al., 1971). By contrast, barnacle geese at the edges of the flocks enjoyed greater access to resources. Edge birds, however, maintained a high vigilance to defend those rich patches against intruders (Black et al., 1992). In other species, food competition intensifies instead in central positions. In impalas, central individuals competed more effectively for food by keeping their head down, which may explain why they were also less vigilant (Blanchard et al., 2008). Another study also suggested more food competition in central positions (Dalmau et al., 2010). Clearly, the relationship between food availability, food competition, and spatial position is crucial to understanding how vigilance should vary across the group (Hirsch, 2007).

Competition for other types of resources can also vary spatially. Competition for mating opportunities, in particular, is often intense. If dominants occupy central positions, peripheral individuals might devote more time to vigilance, especially of the social kind (Alados, 1985). This observation fits with the idea that the need for conspecific monitoring can vary spatially as was the case in common eider broods.

One way to control for many, if not all the confounding factors listed previously, is to experimentally manipulate spatial position in the group. Fernández-Juricic and I controlled for phenotypic attributes and food competition by randomly allocating cowbirds to individual cages placed at different positions in the group (Fernández-Juricic and Beauchamp, 2008). The observed increase in vigilance at the edges of the group could thus be safely attributed to variation in spatial position alone. Another avenue is to focus on groups in which individuals sleep or groom to eliminate issues related to food competition.

4.3.8   Presence of Predators

Pre-emptive vigilance allows individuals to detect predators before it is too late. Prey animals are often acutely aware of the presence or density of predators nearby and should adjust vigilance according to the immediate risk of attack. In a classic study, a trained avian predator was flown over flocks of birds for a few hours each day over several days. In response, birds formed larger groups and invested more time in vigilance (Caraco et al., 1980a). Seeing the predator daily and expecting its presence was clearly sufficient to induce more vigilance.

If predators are nowhere close, it would make sense to reduce investment in vigilance. This assumes, of course, that prey animals can assess the likelihood that predators are nearby. Prey animals could use direct evidence of the presence of predators or indirect cues to assess predator density. Examples of cues provided by terrestrial predators include vocalizations or smells (Kuijper et al., 2014; Nersesian et al., 2012; Périquet et al., 2010). Responses to predator cues have been especially well documented in aquatic habitats where chemical cues transmit easily over large distances (Ferrari et al., 2010). Aquatic predators can be detected directly by odour. In addition, attacks on individuals can leave an odour trail detectable later in the same area providing an indirect cue of predator presence. Unfortunately, whether such direct and indirect cues influence vigilance is not known.

How predator presence influences vigilance has mostly been studied in mammalian herbivore prey species exposed to risk from large predators like lions and wolves. For large herbivores in Africa, vigilance increased as expected when predators were closer (Creel et al., 2014; Périquet et al., 2012; Périquet et al., 2010), which suggests subtle adjustments in behaviour to the immediate risk of predation. A similar finding was documented in elk in response to the presence of nearby wolves (Liley and Creel, 2008; Winnie and Creel, 2007) or in red deer exposed to recent wolf scats (Kuijper et al., 2014). Many other anti-predator responses can be deployed in response to the immediate threat of predation, including changes in group size or habitat type (Caraco et al., 1980a; Creel et al., 2014; Creel and Winnie, 2005; Valeix et al., 2009). Documenting such changes is important because they can potentially confound the effect of predator presence on vigilance.

Not all species investigated thus far adjusted vigilance to the immediate risk of predation (Ciuti et al., 2012; Hall et al., 2013; Mella et al., 2014a; Mella et al., 2014b; Périquet et al., 2012), suggesting that different species weigh the same cues of predation risk differently perhaps reflecting their intrinsic vulnerability to predation or the greater value placed on other cues. Differences between the sexes have also been noted when responding to the presence of predators (Winnie and Creel, 2007), which again suggests that signs of predator presence might be perceived differently by different classes of individuals.

Recent models suggest that vigilance is adjusted to the immediate risk of predation in a dynamic fashion (Beauchamp and Ruxton, 2012a; Sirot and

Pays, 2011). In these models, prey animals pay attention to signs betraying the presence of predators nearby. When the time spent undisturbed increases, the models predict a decrease in vigilance. Downward adjustments in vigilance over time have been documented in some species (Beauchamp and Ruxton, 2012b; Desportes et al., 1991b; Trouilloud et al., 2004; Welp et al., 2004; Wheeler and Hik, 2014). Obviously, other factors can cause a temporal decrease in vigilance, with satiation coming to mind (Section 4.4.2.1). It is thus important to assess alternative explanations when investigating temporal patterns in vigilance.

4.3.9   Environmental Factors

Environmental factors represent extrinsic factors not directly under the control of individuals. Examples include wind speed or light levels. Environmental factors can have two broad consequences for vigilance: they can reduce the ability of prey animals to detect predators or their ability to escape predators after detection. Some environmental factors influence vigilance through changes in physiological requirements. I consider these effects in Section 4.4.

4.3.9.1   Wind

Wind can affect vigilance in many ways. Indirectly, an increase in wind speed can increase energy demands for a predator and lead to more attacks (Willem, 2001). By contrast, impaired manoeuvrability for flying predators on windy days can lead to a reduction in the number of attacks (Quinn and Cresswell, 2004). Changes in perceived predation risk on windy days could lead to upward or downward adjustments in vigilance depending on the expected attack rate. Wind can also carry olfactory information about predators to mammalian prey. Seals at haul out sites, for instance, tend to orient downwind to detect approaching polar bears (Kingsley and Stirling, 1991). Lack of wind, in this case, should increase predation risk and indirectly vigilance.

Wind can also have a more direct effect on vigilance, acting on the prey animals themselves. Windy days can be challenging from the viewpoint of a prey animal visually searching for predators in the vegetation (Carr and Lima, 2010). Moving vegetation makes it harder for prey animals to locate predators using vegetation as a backdrop for their attacks, and could explain higher vigilance on windy days in some species (Blumstein and Daniel, 2003; Carter and Goldizen, 2003; Loughry, 1993; Yasue et al., 2003). This idea could easily be put to the test by experimentally moving vegetation on non-windy days to determine whether such movements really affect vigilance irrespective of other correlated factors like temperature or activity level of predators.

While wind can facilitate olfactory detection of predators, it might, by contrast, degrade or mask auditory signals associated with the approach of a predator. As pointed out earlier, an increase in ambient noise levels has been related to an increase in vigilance in foraging animals (Quinn et al., 2006; Rabin et al., 2006). The effectiveness of collective detection carried out with vocal signals could also be affected if high winds degrade their transmission (Fairbanks and Dobson, 2007; Hayes and Huntly, 2005). This factor alone could lead to an increase in vigilance. The finding that pied babbler sentinels occurred closer to the ground on windy days was interpreted as a means to improve the transmission of vocal signals used to coordinate vigilance (Hollen et al., 2011), suggesting that wind plays an important role in vigilance.

Not all studies reported an effect of wind on vigilance (Dannock et al., 2013; Santema and Clutton-Brock, 2013; Schutz and Schulze, 2011; Shannon et al., 2014a; Valcarcel and Fernández-Juricic, 2009). The effect of wind on vigilance could be speed dependent, and only become apparent at sufficiently high speeds. The range of wind speeds should be provided to assess this possibility.

4.3.9.2 Light Levels

The amount of light should intuitively affect the quality of information gathered visually. Prey animals that detect predators visually might therefore be expected to increase vigilance at low light levels. In line with this prediction, maximum vigilance in dark-eyed juncos occurred in dim light early in the day. Interestingly, avian predators also tended to be more active at this time of day (Lima, 1988). This increase in vigilance early in the day happens at a time when animals are expected to be hungry after fasting all night. Hungry animals should be less likely to invest in anti-predator vigilance as it interferes with the acquisition of resources (McNamara and Houston, 1992). Low light levels must therefore exert a very strong influence on perceived predation risk to mask the effect of hunger.

Cloud cover during the day can substantially decrease light levels, suggesting that vigilance should increase on cloudy days. In two species of birds, vigilance increased as predicted on more cloudy days (Hilton et al., 1999b; Lima, 1988). No effect of cloud cover was documented in another avian species (Schutz and Schulze, 2011).

High light levels should allow prey animals to visually detect predators more easily, leading to a reduction in perceived predation risk (Kotler et al., 2010). Paradoxically, high light levels might instead increase predation risk by making prey animals more visible to their predators (Kacelnik, 1979). Too much light could also interfere with image resolution in the eye (Fernández-Juricic and Tran, 2007). Poor visual resolution could affect the ability to detect predators and visual alarm signals from companions.

To address these issues, researchers investigated whether ambient light levels influence patch preferences and vigilance in prey animals. Two species of birds preferred to forage in shady patches (Carr and Lima, 2014; Fernández-Juricic and Tran, 2007), but small forest passerines preferred more illuminated patches (Carrascal et al., 2001). Foraging in sunlit patches led to a reduction in vigilance in one species (Fernández-Juricic and Tran, 2007) (Fig. 4.8). In another species, light levels did not influence vigilance (Fernández-Juricic et al., 2012). These early, contradictory results probably indicate that light intensity can

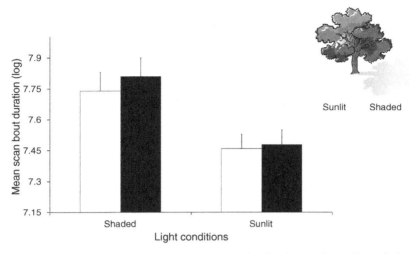

FIGURE 4.8   Light levels and vigilance. Mean scan duration decreased when house finches foraged in sunlit rather than shaded patches, which could affect their ability to detect predators. Foraging patches were located under a tree and could be moved to different locations to take advantage of available shade. The decrease was apparent before (white bars) and after (black bars) a simulated attack. *(Adapted from Fernández-Juricic and Tran (2007).)*

either increase or decrease predation risk. In addition, sunlit patches are not only brighter but also warmer. At low temperatures, such patches might provide warmth to reduce thermoregulatory costs, tilting preferences to brighter patches despite greater exposure to predators (Carr and Lima, 2014).

4.3.9.3   Precipitation

An increase in vigilance on rainy days has been documented in two species (Granquist and Sigurjonsdottir, 2014; Hilton et al., 1999b). It is not clear exactly how precipitation in the form of rain can influence vigilance. Rain could be seen as a visual obstruction, which reduces visibility and indirectly the ability to detect predators. Rain occurs with heavy cloud cover further hindering visibility. Rain also tends to be noisy, which as we saw before, can reduce the effectiveness of predator detection and collective detection. Finally, rain can also increase thermoregulatory costs (McNab, 1980), and affect vigilance indirectly through changes in energy demands. The thermoregulation hypothesis predicts that vigilance decreases on rainy days, which is not compatible with the aforementioned findings. More work is needed to sort through these different mechanisms.

4.3.9.4   Snow Cover

Snow cover on the ground could hamper the ability to escape predators. Elk in Yellowstone National Park can escape from attacking wolves by running to cover, but this is presumably more difficult when snow cover is thicker (Liley and

Creel, 2008). Vigilance in elk did increase with snow cover, but rather weakly. It might be the case that energy demands associated with moving in deep snow actually increased the need to forage at the expense of vigilance. Other hypothetical examples of hampering factors include wind for flying prey animals and difficult terrain for ground-foraging animals.

4.4 DRIVERS OF VIGILANCE ASSOCIATED WITH FOOD OR STATE

When dealing with anti-predator vigilance, researchers naturally focus on drivers associated with predation risk. While avoiding predation is certainly an important component of fitness, animals must also gather sufficient resources to survive. Animals must typically trade-off predation avoidance and foraging efficiency. Any factor that increases the need to acquire resources is thus likely to decrease investment in vigilance. Food availability and energy demands are expected to influence vigilance through this trade-off.

4.4.1 Food Availability

A relationship between vigilance and food availability is expected in two situations. First, animals could use vigilance to locate food patches exploited by nearby conspecifics (Krebs, 1974; Lazarus, 1978; Waite, 1981). This hypothesis predicts a negative relationship between vigilance and current food availability. If this mechanism applies, individuals in poor foraging areas should experience a greater increase in foraging success after using vigilance than searching on their own.

As we saw earlier, the idea that foragers use vigilance to detect foraging opportunities forms the cornerstone of producer-scrounger (PS) models. In a PS system, foragers use the producer mode to locate resources on their own and the scrounger mode to locate resources uncovered by others (Barnard and Sibly, 1981). PS models make several predictions regarding the use of scrounging (Giraldeau and Caraco, 2000), which can be used to predict changes in social vigilance.

The second hypothesis focuses on anti-predator vigilance (McNamara and Houston, 1992). The first model of anti-predator vigilance actually failed to consider food availability at all, and only examined the relationship between vigilance and predator detection (Pulliam, 1973). In a subsequent model, foraging gains were involved explicitly, but foraging could never be terminated abruptly, implying that obtaining food was not constrained by time (Pulliam et al., 1982). Therefore, vigilance was actually independent of food availability. Later, McNamara and Houston (1992) showed that vigilance is expected to decrease with food intake rate when animals face time constraints on foraging. With time constraints, death through starvation becomes a real possibility. The model predicts that sacrifices in safety, through a reduction in vigilance, can be worthwhile when foraging gains, and ultimately fitness, increase rapidly with a decrease in vigilance. A more recent model of vigilance also predicted that vigilance decreases when food is more available (Ale and Brown, 2007).

Some observational studies incidentally related vigilance to food availability (Benhaiem et al., 2008; Cresswell, 1994; Liley and Creel, 2008). However, given that animals often tend to gather in larger groups in areas of greater food availability (Amano et al., 2006; Barnard, 1980; Benhaiem et al., 2008), and that an increase in group size can reduce vigilance on its own, studies that manipulate food availability are more suited to evaluate the relationship between vigilance and food availability.

Experimental studies provided mixed results, with some showing the predicted negative effect of food availability (Butler et al., 2005; Fritz et al., 2002; Pays et al., 2012a; Repasky, 1996), others the opposite effect (Caraco, 1979a; Cézilly and Brun, 1989; Johnson et al., 2001; Randler, 2005a), or simply no effect (Baker et al., 2011; Blumstein et al., 2002; Favreau et al., 2014; Nystrand, 2007; Slotow and Coumi, 2000; Smart et al., 2008; Vahl et al., 2005). In some cases, aggression levels probably increased with either group size or food density (Caraco, 1979a; Johnson et al., 2001), which might have masked the negative effect of food availability on vigilance by increasing the need for social monitoring. Unfortunately, it is not clear whether time constraints prevailed in any of the aforementioned studies.

I examined the effect of food availability on vigilance in fall staging semi palmated sandpipers. Fall staging sandpipers typically double their mass to fuel the long, uninterrupted flight to their over-wintering areas in South America (Hicklin and Gratto-Trevor, 2010). Sandpipers must forage day and night to achieve this goal, indicating that these birds probably face severe time constraints. Any increase in vigilance can potentially decrease food intake rate and reduce migration speed. Controlling for other variables often related to antipredator vigilance, such as distance to obstructive cover and distance to nearest neighbour, I found that vigilance decreased as expected when birds foraged in areas with more food (Beauchamp, 2014a).

For species that combine vigilance and food handling, a positive relationship between food availability and vigilance might simply reflect the fact that when more food is available, proportionately more time is spent handling food rather than searching for it, which would produce an increase in vigilance. To avoid a spurious effect, vigilance should be measured when animals are not handling food.

Many studies of vigilance are observational in nature and rarely control for the availability of resources. Given that food availability can reasonably be expected to influence vigilance, it is important to assess its potential impact on changes in vigilance.

4.4.2 Energy Demands

Energy demands can influence vigilance by altering the rate at which resources are converted to food reserves. When energy demands are high, the balance between foraging gain and predation avoidance should shift in favour of greater resource acquisition at the expense of predation avoidance. In other words, prey animals should be prepared to take greater risk when energy demands are high to maintain their state or reduce the risk of starvation. Therefore, vigilance is expected to decrease when energy demands increase. A caveat is that many environmental factors that affect energy demands can also influence vigilance in a more indirect fashion. Changes in vigilance on a cold or windy day, for example, might be a direct response to increased energy demands, but also indirectly a response to reduced food availability on such days (Yasué et al., 2003). In addition to food availability, increased energy demands have also been associated with changes in group size (Caraco, 1979b; Hogstad, 1988a; Ruckstuhl and Neuhaus, 2009; Shannon et al., 2014a) or level of exposure to risk (Brotons et al., 2000). It is important to consider indirect factors that could confound the effect of energy demands on vigilance.

Energy demands can be influenced by both internal and external factors. Internal factors reflect the internal state of an individual. External factors are factors acting outside the individual like temperature. I focus first on hunger, the best-studied internal factor, and then examine external factors.

4.4.2.1 Hunger

The state of a forager typically refers to the amount of reserves that an animal can draw on for subsistence. An animal with low reserves is expected to increase foraging to reduce the risk of starvation. The amount of reserves available to a forager might not be easy to measure directly, but hunger, which closely matches reserve levels, can easily be quantified through observable behaviour.

A state-dependent model of vigilance incorporates food intake and the amount of reserves available to an animal, which allows us to determine how individuals should adjust vigilance as a function of their state or hunger levels. In their state-dependent model, McNamara and Houston (1992) showed that vigilance generally increases with the amount of available reserves, but the effect varied depending on group size. When group size is small, vigilance is predicted to increase rapidly with the amount of reserves before reaching a plateau. By contrast, when group size is large, vigilance remains low and independent of state until reserves are quite high. In a large group, individuals benefit from predation risk dilution and improved collective detection. They can afford to invest more heavily in foraging at the expense of vigilance to build up reserves, which explains why vigilance is quite insensitive to low reserves levels in large groups. As the level of reserves increases, individuals become less likely to sacrifice safety to reduce the likelihood of starvation, and will invest more in vigilance.

State-dependent vigilance has mostly been explored empirically in sentinel systems. In a sentinel system, one or a few individuals keep a look out for predators from a vantage point while companions forage nearby (Chapter 7). A state-dependent vigilance model adapted to sentinel species pointed out that sentinel behaviour is a low-cost and relatively safe activity only performed when individuals have sufficient reserves (or are less hungry) (Bednekoff, 1997). Sentinels benefit selfishly from this behaviour since it provides maximum security. Others in the group benefit indirectly from this selfish behaviour.

The relationship between sentinel behaviour and reserves has been tested in desert-dwelling Arabian babblers. In a clever experiment, the researchers provided supplemental food to two birds of different social status in several groups and examined whether the increase in body mass (and thus reserves) produced an increase in sentinel effort irrespective of social status (Wright et al., 2001b). As expected from the state-dependent model, sentinel effort increased on days during which targeted individuals received extra food. No such adjustments occurred before or after food provisioning or in other group members (Fig. 4.9). A similar state-dependent response was documented in Florida scrub jays, another sentinel species (Bednekoff and Woolfenden, 2003).

Targeted manipulation of state has also been performed in non-sentinel systems. Food-deprived dark-eyed juncos rejoining their groups showed decreased vigilance, as would be expected from a state-dependent perspective (Lima, 1995a). Consistent with state-dependent vigilance, female Eastern grey kangaroos in poor body condition preferred a vigilant posture that interfered less with foraging (Edwards et al., 2013). Similarly, provisioned female Belding's ground squirrels invested more in vigilance than non-provisioned females (Nunes et al., 2000). In

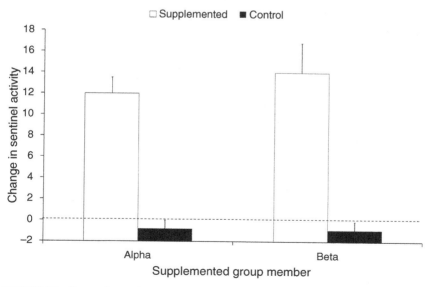

FIGURE 4.9 **Energy demands and vigilance.** In Arabian babblers, the effort allocated to sentinel behaviour increased with satiation. The difference in sentinel activity (amount of time allocated to sentinel behaviour) on supplemented days and on control days was calculated for provisioned birds (supplemented) and control birds (control). Means and standard error bars are shown. *(Adapted from Wright et al. (2001b).)*

this species, food-deprived females were also less likely to interrupt feeding to become vigilant following an alarm call (Bachman, 1993). Food-provisioned ruddy turnstones, a small shorebird foraging on rocky shores, also showed increased vigilance (Beale and Monaghan, 2004a). More testing of the predictions from state-dependent models is needed. Given that the state of foragers might vary substantially amongst individuals in the same group, this could explain why marked individual variation in vigilance occurs within a group.

Relating anti-predator behaviour to time of day provides an indirect way of examining the relationship between state and vigilance. For animals that sleep at night, hunger levels should be high in the early morning and progressively decrease later on. A pattern of increase in vigilance throughout the day should parallel the gradual accumulation of reserves. Such a pattern was documented in a small passerine bird (Pravosudov and Grubb, 1998) and several waterbirds (Burger and Gochfeld, 1998), but more often than not vigilance showed no relationship with time of day (Burger et al., 2000; Gauthier-Clerc et al., 1998; Gauthier and Tardif, 1991; Magle and Angeloni, 2011; Martella et al., 1995; Robinette and Ha, 2001; Rolando et al., 2001; Ward and Low, 1997), or actually decreased (Foster-McDonald et al., 2006; Lima, 1988; Loughry, 1993). Many factors, unfortunately, also vary with time of day, making it difficult to interpret the results. In the dark-eyed junco study, for instance, early morning was not only associated with greater hunger levels but also with dim light and more active predators. Time of day can also correlate with temperature (Petrie and Petrie, 1998), prey availability (Rolando et al., 2001) or predator activity (Loughry, 1993; Ward and Low, 1997), which can all influence vigilance on their own.

4.4.2.2 Temperature

Temperature is the environmental factor that has received the most attention in the context of energy demands. Some studies showed the expected increase in vigilance as temperature rises (and energy demands ease) (Beveridge and Deag, 1987; Carter and Goldizen, 2003; Pravosudov and Grubb, 1995, 1998), but others did not (Boysen et al., 2001; Jónsson and Afton, 2009; Robinette and Ha, 2001; Ruckstuhl and Neuhaus, 2009; Shannon et al., 2014a). Some of the best evidence for a relationship between temperature and vigilance comes from a study where the effect of temperature could be isolated from the effect of group size and food availability (Pravosudov and Grubb, 1995) (Fig. 4.10). One possibility to explain the aforementioned negative findings is that animals in these studies faced little time constraints on foraging. A decrease in temperature may not necessarily lead to a decrease in vigilance if individuals compensate for higher energy demands by increasing time spent foraging instead (Boysen et al., 2001) or cutting on time spent resting or grooming.

The possibility that animals compensate for increased energy demands by foraging more can be safely discarded by studying how temperature affects vigilance when animals are not foraging. Such a study was conducted in sleeping black-tailed godwits, a large shorebird species that forms tight aggregations when

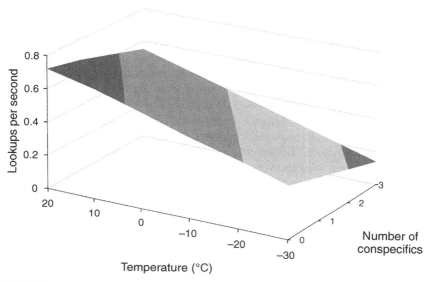

FIGURE 4.10   Temperature and vigilance. As temperature decreased, the number of lookups per second, a measure of vigilance, decreased in tufted titmice controlling for the number of conspecifics in the group. The plane illustrates the best-fitted equation relating vigilance to the number of conspecifics and temperature. *(Adapted from Pravosudov and Grubb (1995).)*

sleeping. Peeking during sleep, the measure of vigilance, surprisingly increased with rising temperatures (Dominguez and Vidal, 2007). This finding suggests that the very act of vigilance itself carries energy costs. Less vigilance during sleep keeps the head closer to the body and perhaps reduced energy expenditure at low temperatures. Another possibility is that raising the head during vigilance acts as a cooling mechanism in hot weather (Dominguez and Vidal, 2007). Whether such responses also apply when animals are foraging is not known.

Temperature could influence social vigilance as well. Warmer temperatures might stimulate reproduction, inciting animals to spend more time monitoring rivals and mates (Beveridge and Deag, 1987). In Northwestern crows, vigilance actually increased at low temperatures (Robinette and Ha, 2001). In this species, vigilance is partly aimed at monitoring conspecifics for opportunities to steal food. It might be the case that when temperature is low food is more difficult to find. Scrounging would become more prevalent at low temperatures with a concomitant increase in social vigilance.

4.5   INDIVIDUAL DIFFERENCES

The previous sections amply demonstrated that vigilance can vary consistently amongst individuals in relation to factors like sex, age or dominance status. Further differences within a particular class of individuals have been related to individual differences in personality traits, such as shyness or boldness (Wilson et al., 1994).

To invoke personality traits, individual vigilance must be stable across conditions and over time. Few studies have reported individual vigilance profiles. In a laboratory study with nutmeg mannikins, some birds were consistently more vigilant than others across a range of group sizes (Rieucau et al., 2010), but stability over time was not assessed. In the field, redshanks showed consistent individual differences in vigilance across two different locations over a period of several weeks (Couchoux and Cresswell, 2012), suggesting the existence of stable inter-individual differences in vigilance. Another shorebird species showed consistent inter-individual differences in vigilance across three different levels of predation risk (Mathot et al., 2011). Adult spotted hyaenas also displayed consistent patterns of vigilance within but not across contexts (Pangle and Holekamp, 2010b). In white-tailed prairies dogs, where individual variation in vigilance can be quite high, half of the individuals maintained a similar level of vigilance over two successive years (Hoogland et al., 2013). Significant amongst-individual variation in vigilance has also been documented recently in nesting cliff swallows (Roche and Brown, 2013).

Consistency in vigilance has been the most studied in Eastern grey kangaroos. Initially, research showed that group size in this species affected vigilance to different extent in different individuals (Carter et al., 2009). A further study documented consistent patterns of vigilance in a number of females (Dannock et al., 2013). In addition, females that fled earlier from an approaching threat also tended to be more vigilant (Edwards et al., 2013), suggesting a link between boldness and vigilance. In this species, about 7% of the total time spent vigilant was accounted for by individual variation, a non-trivial, albeit small source of variation. When tested across different ecological conditions, individuals maintained consistent differences in vigilance, but showed the same level of flexibility in adjustments (Favreau et al., 2014) (Fig. 4.11). Overall, these results strongly suggest consistent inter-individual differences in vigilance.

The consequences of inter-individual differences in vigilance have been little studied. In cliff swallows, there was no relationship between the level of vigilance and reproductive success. In fallow deer, boldness was associated with vigilance, and the researchers speculated that bolder individuals and their offspring might experience a higher predation risk (Bergvall et al., 2011). A similar association between predation risk and boldness was suggested in a species of lizard (Carter et al., 2010). Individuals that invest little in vigilance will fare better in a group composed of vigilant rather than less vigilant conspecifics, suggesting that individual fitness may vary depending on group composition. Whether individual vigilance profiles influence group composition is not known. Stable inter-individual differences might also explain the wide scatter in plots of the relationship between vigilance and group size displayed in many species (Beauchamp, 2013b).

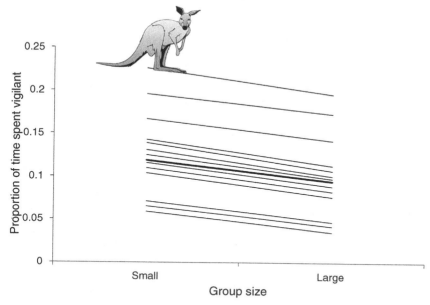

FIGURE 4.11 Stable inter-individual patterns of vigilance. In Eastern grey kangaroos, vigilance decreased with group size to the same extent in all individuals, but different individuals maintain different levels of overall vigilance. The thick line shows the average in the population. Each line represents a different female kangaroo, but only a sub-set of all individuals is shown here. *(Adapted from Favreau et al. (2014).)*

Documenting inter-individual patterns of vigilance faces several challenges. Not least amongst them is over how many conditions and across how much time should vigilance be consistent before invoking personality traits. In addition, inter-individual differences might be more difficult to discern in some contexts than others. In a large group, for instance, where the level of competition for resources can be intense, vigilance may be uniformly low to increase foraging efficiency.

Consistent inter-individual patterns of vigilance may betray personality traits. However, they can also reflect consistent individual differences in state. For instance, an individual that is consistently underweight or food-deprived would show low vigilance levels across different conditions and over time. Rather than indicate a bold, risk-prone personality, consistently low vigilance would reflect poor foraging efficiency instead.

4.6 CONCLUSIONS

When discussing various drivers of animal vigilance, I painted a relatively black and white picture of their effects: empirical patterns simply fitted the expectation or not. A proper meta-analysis of the vast literature devoted to most drivers of vigilance remains to be done. In a meta-analysis, findings are culled from several studies and quantified using standard measures of effect sizes like the correlation coefficient (Arnqvist and Wooster, 1995; Nakagawa and Santos, 2012). Such estimates can also be weighted to put more emphasis on arguably more reliable estimates. The overall effect size is obtained after pooling all the available information. I think it would be much more enlightening to conclude that sex explains 12% of the variation in vigilance on average in 20 studies than to say that sex had a statistically significant effect on vigilance in seven of these studies. A more quantitative approach may also shed light on variation in the strength of the relationship between vigilance and a particular driver in different sub-sets of a population or under different ecological situations.

I want to raise another issue regarding the drivers of vigilance. While many of the drivers of vigilance can act alone, joint effects are probably quite common. Distance to cover, as an example, can covary with food availability, group size or dominance status. Three options are available to isolate the effect of one particular driver. First, researchers can narrow the set of observations included in statistical analyses. In Eastern grey kangaroos, researchers thus focused on peripheral females in groups exploiting relatively homogeneous food (Favreau et al., 2014), which ruled out the effect of sex, position in the group, and food availability.

The second option manipulates one factor of interest while controlling for other drivers of vigilance. In our study of brown-headed cowbirds, Fernández-Juricic and I manipulated group density by moving individual cages at different distances from one another. Food availability was similar in each cage, and we included sex and body condition of the foragers as co-factors in the statistical analysis. Obviously, not all study species are amenable to this kind of controlled experiments. In addition, care must be taken to provide conditions that mimic those found in the field.

The third option is to include many potential drivers of vigilance in a multi-variable statistical model, and to report the effect of a particular driver of vigilance holding the effect of the other drivers constant. Liley and Creel (2008) used this approach to examine the contribution of several different types of variables to vigilance in elk. Drivers of vigilance included factors associated with prey (e.g. sex or group size of elk), predators (e.g. presence of wolves) or the environment (e.g. snow cover). As long as these various factors are not hopelessly correlated with one another, the multi-variable statistical approach can be used to identify factors that are consistently found in the models that best fit the data. In elk, some prey and predator factors best explained vigilance.

The meta-analysis approach that I suggested earlier may come in handy for the selection of variables to be considered in future analyses of vigilance. Drivers of vigilance that explain relatively little variation on average may be ignored at no great loss while those that are relatively more important should be controlled in one of the three ways described previously.

Chapter 5

Animal Vigilance and Group Size: Theory

5.1 INTRODUCTION

Of all the drivers of vigilance, group size has attracted the most attention. The size of a group is relatively easy to measure in the field and can be manipulated in the laboratory. All models predict that individual anti-predator vigilance decreases in larger groups. For all these reasons, group size has become over the years an obvious target for a research project on vigilance. Adding to the appeal, the expected decrease in vigilance with group size appears rather counterintuitive at first sight. Because the purpose of anti-predator vigilance is to detect predators, any reduction in vigilance would seem to imply that each individual in a group is more, rather than less, vulnerable. However, research has clearly shown that living in groups can reduce predation risk, making the decrease in individual vigilance the expected solution to maximize fitness as group size increases. In this chapter, I explore the theoretical underpinnings of the group-size effect on vigilance. In the following chapter, I provide a review of the empirical evidence supporting this effect in birds and mammals.

5.2 PREDATION RISK AND GROUP SIZE

Several mechanisms, which I review further, can decrease predation risk for individual group members. These mechanisms can act alone or in combination to influence predation risk. Models of anti-predator vigilance include at least one or typically several of these mechanisms.

5.2.1 Many-Eyes Effect

Speculating about the function of mixed-species flocking in Amazonian birds, Bates, a British naturalist from the nineteenth century, surmised that living in groups reduces the chances a predator can approach flock members undetected (Bates, 1863). This idea was also articulated by other avian researchers (Belt, 1874; Miller, 1922). Galton (1871) working with cattle in Africa at about the same time, also noted that in a herd there are always some ears or noses available to detect an approaching threat. Being in a group thus multiplies the senses dedicated to predator detection and should allow earlier detection of threats. The

Animal Vigilance. http://dx.doi.org/10.1016/B978-0-12-801983-2.00005-X

117

multiplicity of senses argument is typically referred to as the many-eyes effect (Caraco et al., 1980b; Sigg, 1980), although it is clear that other senses can be involved in predator detection. Increased predator detection through the many-eyes effect lies at the core of all models of anti-predator vigilance including group size.

5.2.2 Collective Detection

The concept of collective detection, also known as mutual warning, goes back to Galton who noted that in a herd of cattle an alarmed individual soon warns all the others about danger. Mutual warning allows all group members, including those that failed to detect the predator early, to escape before the predator gets fatally close (Lima, 1990). Information transfer from threat detectors to non-detectors defines collective detection. The impact of collective detection can be measured, theoretically at least, as the difference between the probability of surviving an attack for a non-detector when at least one other group member has detected the approaching predator, and the probability of surviving for the same non-detector when no other group member has detected the predator. It is clear that without collective detection, early detection of threats in a group through the many-eyes effect can only benefit early detectors.

Collective detection can involve many types of cues. Acoustic cues, in particular, have many desirable properties for information transfer about threats. They can be loud and carry over large distances. Examples of acoustic cues include not only signals intended to alert other group members, such as alarm calls (Hoogland, 1979; van der Veen, 2002) but also indirect cues associated with escape. The escape plop produced by Iberian green frogs jumping into the water indirectly warns nearby conspecifics about imminent danger (Martín et al., 2006). Whistles or whirring sounds produced by rapid take-off in birds can also transfer information about threats very quickly (Coleman, 2008; Hingee and Magrath, 2009). Highlighting the role of indirect acoustic cues of alarm, common voles prevented from hearing escape noises from alarmed companions reacted less often to a threat (Gerkema and Verhulst, 1990). Sounds of alarm can transfer information about threats, but in some cases the absence of acoustic cues may also reliably signal alarm. Indeed, alarmed individuals often cease normal activity to investigate a potential threat. The ensuing silence has been shown to act as an indirect signal of threat in rats (Pereira et al., 2012) (Fig. 5.1).

Visual cues tend to have a shorter range than acoustic cues, and can only be perceived when the receiver faces the sender. Examples of visual cues used in collective detection include alert postures (Brown et al., 1999a; Magurran and Higham, 1988) and rapid movements in response to threats (Berger, 1978; Caro, 1986; Handegard et al., 2012; Treherne and Foster, 1981). While flight by only one group member may be ambiguous, several flight reactions in rapid succession can provide a stronger alarm signal (Cresswell et al., 2000; Lima, 1995b; Marras et al., 2012). In some species, mechanical signals also play an important role in collective detection. Fright or flight reactions in the

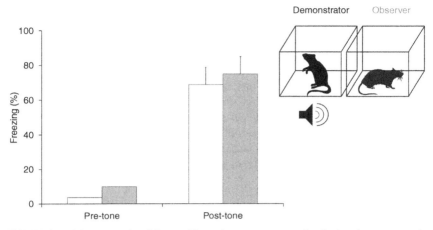

FIGURE 5.1   Silence can signal danger. Naïve observer rats exposed to the freezing response of a demonstrator rat, which was conditioned by a sound signal, showed freezing responses to a greater extent after than before hearing the signal. Further experiments showed that the mere absence of sound from the demonstrator's cage was sufficient to trigger a freezing response in observer rats. Means and standard error bars are shown. Demonstrator in white and observer in grey. *(Adapted from Pereira et al. (2012).)*

water transmit as waves readily detectable by nearby individuals (Vulinec and Miller, 1989). Foot-drumming mammals transmit alarm through vibration in the ground detectable by individuals out of visual or auditory range (Randall, 2001).

Regardless of the medium used for information transfer, effective collective detection cues enable non-detectors to escape more rapidly than would be the case without signals or cues from detectors. Effectiveness of collective detection can be measured by the speed of alarm transmission within the group. In a marine insect, for instance, fright movements by foragers at the edges of the group propagated so rapidly through the group that non-detectors initiated escape behaviour before the predator was in sight (Treherne and Foster, 1981). Effective collective detection can also be inferred if non-detectors in a group respond more quickly than if they were alone or did not receive the warning signal (Fernández-Juricic et al., 2009; Martín et al., 2006; Webb, 1982). On this basis, some bird species showed ineffective collective detection perhaps because fright reactions in birds tend to be more ambiguous (Fernández-Juricic et al., 2009; Lima, 1995b; Roth et al., 2008b).

Collective detection relies on the transmission of visual, auditory or mechanical cues of alarm to other group members. Therefore, any environmental factor that interferes with information transfer should influence the effectiveness of collective detection. Visual barriers between foraging conspecifics is a good case in point. In one study with small birds foraging on the ground, non-detectors reacted less strongly to alarm when a small wall prevented individuals from seeing the early stages of flushes to safety (Lima and Zollner, 1996).

The distance between group members can also influence collective detection. At some point, auditory or mechanical cues provided by distant neighbours will become imperceptible or neighbours will be too difficult to monitor visually, making collective detection less likely. Consistent with this expectation, the effectiveness of collective detection decreased with distance in common voles (Gerkema and Verhulst, 1990), in Eastern grey kangaroos (Pays et al., 2013), and in small passerine birds (Lima and Zollner, 1996). Recent work with birds showed that the probability of detecting a flushing response decreased rapidly with distance (Fernández-Juricic and Kowalski, 2011), limiting collective detection to a narrow range about each individual. The effectiveness of collective detection often forms an integral part of anti-predator vigilance models.

5.2.3 Dilution

Dilution acts after the launch of an attack and operates whether or not the predator has been detected. If the predator cannot capture all group members after an attack, the individual risk of being targeted will decrease proportionately to the size of the group (Bertram, 1978). Risk dilution works due to the presence of many alternative targets in the group for the predator. Finding that the rate of attack fails to increase with group size is not sufficient to infer that individual attack rate will be equal to $1/N$ (the dilution effect), where N represents group size. This is because some individuals in the group are often more likely to be attacked than others. This is the case notably for peripheral individuals in a group (Chapter 4). Risk dilution will reduce the risk of attack per individual as long as the risk of attack at the group level does not increase proportionately faster. If the predator targets slow reacting group members during an attack, the pool of individuals available to dilute risk will be much lower than the size of the group (Bednekoff and Lima, 1998b).

There is much evidence supporting a decrease in the proportion of individuals captured per unit time during an attack as group size increases (Foster and Treherne, 1981; Jensen and Larsson, 2002; Lucas and Brodeur, 2001; Morgan and Godin, 1985; Sorato et al., 2012; Turchin and Kareiva, 1989; Uetz and Hieber, 1994; Wrona and Dixon, 1991). However, there are alternative explanations for a decrease in *per capita* attack rate in larger groups. For instance, larger groups might be more efficient at detecting or avoiding predators, or might confuse them (Beauchamp, 2014b).

Dilution and collective detection often make the same predictions regarding anti-predator behaviour, and their relative contributions can be difficult to disentangle. As an example, a reduction in individual anti-predator vigilance in a group can arise because predation risk is diluted and/or because collective detection is efficient (Roberts, 1996). To disentangle these effects, Roberts (1996) suggested studying anti-predator vigilance in settings involving just one mechanism. To this end, Ruxton and I investigated anti-predator behaviour in flocks of semipalmated sandpipers foraging on open mudflats (Beauchamp and Ruxton, 2008). Peregrine falcons, the main predator of these small shorebirds during migratory stopover, launch their attacks from the wooded area bordering mudflats. As all attacks originate from this general direction, the side of the flock facing cover is riskier for sandpipers. Collective detection can safely be assumed constant because no obstacles prevent mutual warning in the open mudflat habitat. Sandpipers on the riskier side of the flock are first in line in case of attack, and can only dilute predation risk with neighbours on that side. On the opposite side of the flock, individuals benefit from the presence of several layers of protection. We expected that individuals on the riskier side of the flocks would invest more in anti-predator defences, such as vigilance, at the expense of foraging. As expected, a lower foraging intensity prevailed on the riskier side of the flocks (Fig. 5.2). We concluded that risk dilution acting alone produced these differences in foraging intensity.

Another approach to isolating the effect of risk dilution focuses on vigilance in mixed-species groups, groups that contain several species each of which can face a unique set of predators. In mixed-species groups, some species contribute to the detection effect but not necessarily to risk dilution because different prey species need not share the same predators (Schmitt et al., 2014). The expected contribution from risk dilution can then be related to vigilance in groups with different species composition (Chapter 8). Other researchers have used a statistical approach to evaluate the relative contributions of risk dilution and collective detection to vigilance (Dehn, 1990; Fairbanks and Dobson, 2007; Rieucau et al., 2012).

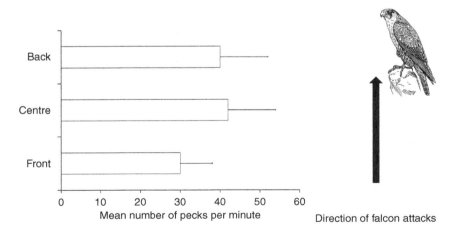

FIGURE 5.2 Disentangling collective detection and risk dilution. Semipalmated sandpipers foraging on the riskier side of flocks tended to peck at a lower rate than those on the opposite side. The lower peck rate suggested a larger investment in anti-predator vigilance. The riskier side of a flock faced the area from which peregrine falcons launched their attacks. The number of sandpipers available to dilute predation risk increases when moving away from the riskier side of the flock while collective detection is assumed constant across the flock. Means and standard deviation bars are shown. *(Adapted from Beauchamp and Ruxton (2008).)*

To model dilution, predation risk is typically divided equally amongst all group members. However, as suggested earlier, some individuals in the group are often more at risk than others. Phenotypic attributes that bias predation risk within a group are important to consider when modelling the benefits of dilution (Dehn, 1990; Rieucau et al., 2012).

5.2.4 Confusion

Confusion acts when prey capture is imminent. Confusion can be invoked when attack success decreases with group size. Naturalists noted early that the scattering of group members following an attack can reduce capture rate by dividing the attention of the predator (Allen, 1920; Grinnell, 1903; Landeau and Terborgh, 1986; Lavalli and Spanier, 2001; Miller, 1922; Neill and Cullen, 1974). In a particularly telling example, predatory bass always captured minnows presented singly, but failed most of the times when their prey occurred in large groups (Landeau and Terborgh, 1986).

Confusion exploits the cognitive limitations of the predator by reducing the ability of the predator to select and/or pursue a target in the fleeing group (Krakauer, 1995). The observation that predatory three-spined sticklebacks prefer to attack stragglers in a swarm of *Daphnia* (Milinski, 1977) was attributed to confusion. Further experiments with this predator–prey system showed that many factors reduced the number of predation attempts per individual prey, including higher movement speed, movements in the same direction, increased swarm size and density, and increased uniformity in the appearance of swarm members (Ohguchi, 1981), all of which can potentially increase the confusion effect.

5.3 MODELS OF ANIMAL VIGILANCE IN GROUPS

Models of anti-predator vigilance in animals combine the various factors described earlier to predict how vigilance should vary with group size. Testing predictions from these models has been a mainstay of vigilance research over the last 40 years (Chapter 6). Here, I describe various models to highlight their assumptions and predictions.

5.3.1 Early Models

The first model of anti-predator vigilance in relation to group size was proposed in the early 1970s (Pulliam, 1973). Pulliam envisaged a situation in which a predator attacks a group of prey animals at random times. The predator needs time T from the beginning of the attack to get close enough to the group to make a capture. Prey animals in the group alternate between feeding and vigilance. An approaching predator can only be detected during vigilance, but not while feeding. All scans to detect a predator are short and constant in duration. Recall that the length of time between two successive scans is called an inter-scan interval.

The key assumption in the model is that a prey animal in the group interrupts its feeding to scan at random times dictated by a Poisson process with fixed rate λ. With the Poisson scan initiation process, the distribution of inter-scan intervals follows a negative exponential distribution. The fixed rate assumption in the Poisson process implies that the initiation of a scan is independent of the current inter-scan interval duration. In addition, scanning interruptions occur independently amongst individuals in the same group since they are dictated by the same but independent Poisson processes.

Consider first the situation in which the predator attacks a solitary forager. The attack will be successful if the forager fails to scan before time T, an event which occurs with probability $e^{-\lambda T}$. Now, consider the case where the predator targets one individual in a group of N individuals. If any group member initiates a scan before time T, all individuals can escape to safety, reflecting perfect collective detection. A feeding group member is in danger if no one in the group scans during time T. Assuming independent scanning by all group members, the situation where no one scans before time T arises with probability $e^{-N\lambda T}$. Therefore, the probability of detection by at least one individual in the group is the complement, namely, $1-e^{N\lambda T}$. This value represents the probability of surviving an attack.

Naturally, the probability of detection for a solitary forager increases when the predator takes longer to attack. However, gains in the probability of detection become progressively smaller as T increases because the chances of not scanning become increasingly small as attack time gets longer (Fig. 5.3). The probability of detection changes similarly with attack time for a group member. But, for any attack time, the probability of detection increases with group size. When groups are large enough, the probability of detection gets very close to 1. In this case, predator detection is nearly unavoidable, and the chances of capture are very small (Fig. 5.3).

When all the senses are pooled, individual group members are well protected from attacks and cannot easily be surprised by predators. Pulliam showed that group members could actually reduce their rate of scanning and still enjoy a survival advantage over solitary foragers. Essentially, Pulliam showed mathematically what Lack (1954) and Galton (1871) intuited earlier, namely, that because living in a group provides so much safety, each group member could maintain a lower level of vigilance than a solitary forager. The model thus makes two testable predictions: (1) the probability of detection increases with group size, and (2) individual investment in vigilance decreases with group size.

Pulliam's 1973 model focused only on predator detection. However, it is clear that survival depends not only on predation avoidance, but also on accumulating sufficient reserves to avoid starvation. In Pulliam's model, there are no costs to bring vigilance down. Subsequent models of anti-predator vigilance all include a trade-off between vigilance and foraging. With a trade-off, achieving greater safety comes at the expense of feeding. The many-eyes effect, the mechanism modelled by Pulliam, is but one of several factors that can reduce

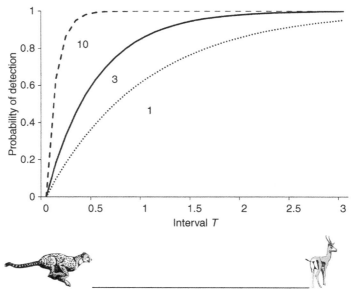

FIGURE 5.3 Probability of detection and attack duration. The probability of detection increases but at a decreasing rate as the time needed to mount an attack increases (interval T). This probability also increases with group size for any time T. The probability of detection is based on the assumption that scans are initiated following a Poisson process. Any scan initiated before time T allows a prey animal to escape successfully. This interval is illustrated for a cheetah launching a surprise attack on a gazelle.

predation risk in a group. Other factors like encounter/dilution effects and confusion were included in subsequent models.

With hindsight, some assumptions in the model are quite questionable. For instance, Pulliam did not justify the assumption of random scanning. It is easy to imagine other patterns of scan initiation, including interruptions at regular intervals, which may actually be more beneficial to foragers (Bednekoff and Lima, 2002; Scannell et al., 2001). The model also assumes that attack rate is independent of group size, which is not the case in some predator–prey systems (Cresswell, 1994; Hebblewhite and Pletscher, 2002).

Pulliam's model also ignores the issue of cheating. When safety is traded-off against foraging gains, an individual that unilaterally reduced its own vigilance would get the benefit of increased detection provided by the remaining group members without paying the full cost of investing time in vigilance. Time savings could be allocated to other fitness-enhancing activities such as resting or foraging. In response to a unilateral defection, or cheating, the only solution for other group members is to decrease their own vigilance (discussed later). Modelling anti-predator vigilance requires an approach that takes cheating into account.

Pulliam et al. (1982) proposed a new model of anti-predator vigilance some 10 years later. This model addressed many of the shortcomings detailed previously. In the model, living in a group still improves the chances of detecting a predator through the many-eyes effect and perfect collective detection. But time spent vigilant now comes at a cost because the time taken by vigilance reduces the rate at which food can be gathered. With a lower food intake rate, individuals must forage longer and be exposed to more attacks by predators to achieve the same overall intake at the end of the day. In other words, the model incorporates a trade-off between safety and foraging gains. The total number of attacks experienced by a forager during the day determines survival, but because all individuals forage until a fixed food requirement is met, starvation is not an issue for survival.

To incorporate cheating, the model was formulated as a game between group members. If cheating is possible, the solution that maximizes success for all group members may not prove stable because a defector, using a different pattern of vigilance, might enjoy greater success than the others. For instance, a less vigilant cheater would get more food but enjoy the same safety as the others.

Two possible solutions emerged in this vigilance game. The vigilance level that maximizes fitness at the group level represents the optimal solution. This solution, however, can easily be invaded by cheaters. The evolutionary stable, or selfish solution, arises when no group member choosing an alternative level of vigilance can achieve a higher fitness. Observed scanning rates in groups of different sizes were compared to those predicted under the two types of solutions in yellow-eyed juncos, a small granivorous bird species (Fig. 5.4). In

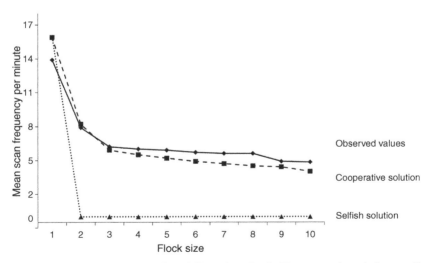

FIGURE 5.4 **Cooperative or selfish vigilance in animals.** The cooperative solution specifies the level of vigilance that maximizes fitness at each flock size. The selfish solution represents the evolutionary stable solution: no deviation from the selfish solution, when it is adopted by all flock members, can provide a higher fitness. In yellow-eyed juncos, the selfish solution proved a worst fit than the cooperative solution to explain scanning rates. *(Adapted from Pulliam et al. (1982).)*

this species, the optimal, or cooperative solution, provided a better fit to the observed scanning rates than the selfish solution. The selfish solution identified by the model specifies no vigilance at all in groups. This solution might be too extreme. If vigilance provided direct benefits to individuals during an attack, say, by allowing them to escape sooner than non-detectors, it seems unlikely that vigilance would fall to such low levels.

5.3.2 McNamara and Houston's 1992 Model

This sophisticated model incorporated several new elements in anti-predator vigilance models (McNamara and Houston, 1992). The first is the possibility that direct detection of a threat, rather than indirect detection through collective detection, can increase the chances of surviving an attack. Detection of a threat through one's own vigilance can give a head start when escaping, thus increasing survival prospects following an attack (Packer and Abrams, 1990) (Box 5.1).

BOX 5.1 McNamara and Houston's 1992 Model

I present a condensed version to illustrate key concepts in the model. The group consists of N foragers, each of which alternates between feeding and vigilance. Scans are assumed to be short and constant in duration τ. For group member i, a bout of feeding lasts for a random time drawn from an exponential distribution with mean λ^{-1}. This determines the length of an inter-scan interval and thus the rate at which scans are initiated. The expected proportion of time spent feeding (U_i) for this individual is given by:

$$U_i = \frac{\frac{1}{\lambda_i}}{\tau + \frac{1}{\lambda_i}} = \frac{1}{1 + \lambda_i \tau}$$

Now, I turn to the attack scenario. As was the case in Pulliam's 1973 model, a predator attacks the group at random times and must go undetected for time T to be successful. If the proportion of time spent feeding is denoted by u, the probability that one group member fails to detect the predator during this time is:

$$g(u) = u \exp\left(-\frac{T}{\tau}\left(\frac{1}{u} - 1\right)\right)$$

The exponential term is equivalent to $e^{-\lambda T}$ in Pulliam's model. Knowing the probability of non-detection for a single forager, it is possible to calculate the probability D_i that this focal group member will die after an attack:

$$D_i = P_n Q_n + P_{nf} Q_{nf} + P_f Q_f$$

This probability consists of three terms: the probability P_n of no group members detecting the predator, the probability P_{nf} of at least one other group member but not the focal individual detecting the predator, and, finally, the probability P_f of the focal individual detecting the predator on its own. The probability of death following an attack for the focal individual where either no group member detects the predator, or one or more individuals not including the focal individual detects the predator, or the focal individual itself detects the predator are Q_n, Q_{nf} and Q_f, respectively.

Assuming that the focal individual adopts strategy u_i (the proportion of time devoted to feeding) and that all other foragers in the group adopt strategies u_j ($j > 1$), the probability that no group member detects the predator is the product of the probabilities that all individuals fail to detect the predator:

$$P_n = g(u_i) \prod_{j=1, j \neq i}^{N} g(u_j)$$

To calculate P_{nf}, we need the probability that all the non-focal individuals fail to detect the predator. When subtracting this from one, we get the probability that at least one of the non-focal group members detects the predator. This is then multiplied by the probability that the focal individual i has not detected the predator

$$P_{nf} = g(u_i) \left(1 - \prod_{j=1, j \neq i}^{N} g(u_j) \right)$$

P_f is simply the complement of the probability of not detecting the predator, namely,

$$P_f = 1 - g(u_i)$$

Now I turn to the selection strategy for the predator when attacking the group, which determines the probability of death upon attack. The predator is assumed to select a target from each of the N group members randomly, which means that each individual has a probability $1/N$ of being selected. This is simply the risk dilution factor. If no group member detects the predator then obviously the targeted forager is successfully captured and the probability of death for the focal forager is $1/N$. If detection by at least one group member occurs, the predator will experience reduced success. When relying on collective detection, which means that the focal forager has not detected the predator directly but at least one companion did, the probability of being targeted is lower than if no one has detected the predator because of the head start provided by collective detection. Essentially, the probability of selection is lower with collective detection. This is lowered to an even greater extent if the focal forager detects the predator directly rather than through collective detection. With these assumptions in mind, we have:

$$Q_n = \frac{1}{N},$$

$$Q_{nf} = \frac{b}{N},$$

(Continued)

and

$$Q_f = \frac{c}{N}.$$

Where $c < b < 1$.

When $b = 1$, collective detection fails to operate, but becomes perfect when $b = c$. When b is much greater than c, the benefit of detection through one's own vigilance is much higher, which provides a way to model the advantage to detectors. There is no advantage to detectors when $b = c$ (Fig. 5.5).

The focal forager i must gather an amount of food F before leaving the group for safety, and food is gathered at rate ϕ whenever it is feeding. This means that this forager will be exposed to predator attacks for time $F/(\phi u_i)$. If predators attack randomly at rate α the probability of focal individual i surviving is

$$L = \exp\left(-\frac{\alpha F D_i}{\phi u_i}\right)$$

This was the easy part. McNamara and Houston then provided a way to find the stable solution to this system of equations such that no alternative strategy for investing in vigilance provided a higher fitness to cheaters.

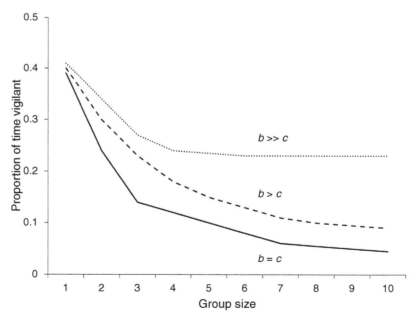

FIGURE 5.5   Advantage to detectors. In McNamara and Houston (1992) anti-predator vigilance model, the proportion of time spent vigilant increases when detectors fare better than non-detectors after an attack. Detectors enjoy an advantage over non-detectors because they can flee earlier. The advantage to detectors depends on the relative value of two model parameters: when b is larger than c the detector is relatively less likely to die upon attack (see text for details).

As with all models of vigilance including a trade-off between predation avoidance and foraging efficiency, the model predicted a decrease in the evolutionary stable level of vigilance (the selfish solution) as group size increases. However, the magnitude of the group-size effect on vigilance depends on the relative advantage of direct versus indirect detection. When there is an advantage of detecting threats directly rather than indirectly, the model predicted higher vigilance at each group size and a shallower decrease in vigilance with group size. This personal advantage of using vigilance implies a greater reliance on one's own vigilance to detect threats. This finding also highlights the fact that rather than being a fixed property, the magnitude of the group-size effect on vigilance can be modulated.

The model also considered the confusion effect. As argued earlier, the hurried departure of several individuals in a group can confuse the predator. McNamara and Houston incorporated the confusion effect in a particular version of their model by reducing the probability of death upon attack in larger groups. With the confusion effect, the stable level of vigilance decreased more slowly with group size. This appears at first sight a rather unexpected outcome. Confusion works by providing extra safety and should allow individuals to decrease their vigilance in larger groups. McNamara and Houston investigated the effect of confusion in a model where detectors and non-detectors alike benefit to the same extent from confusion. Because confusion only works in this model when the predator is detected, an increase in vigilance allows foragers to increase the chances of detection, and confusion then kicks in to reduce the risk of capture following the attack. While this increase in vigilance will expose the foragers to more attacks, the confusion effect allows them to reduce the effectiveness of each attack.

McNamara and Houston pointed out that previous models of anti-predator vigilance implied that vigilance is independent of predation risk (Lima, 1987b). This is an odd situation given that predation risk should be the very factor driving anti-predator vigilance in the first place. In the Pulliam et al. (1982) model, for instance, predation risk played no part in the stable solution because foragers faced no risk of starvation. In McNamara and Houston's approach, foraging can end prematurely due to external disturbances, which means that starvation is likely if the rate of food intake is too low. When there is a risk of starvation, the stable level of vigilance becomes sensitive to predation risk with the obvious prediction that vigilance increases when attacks are more frequent.

Finally, McNamara and Houston included the level of energy reserves available to a forager. Internal reserves can be drawn upon to sustain a forager and must be maintained at a sufficient level to avoid starvation. Models that include reserve levels are said to be state dependent. Naturally, a forager with more reserves should be less likely to sacrifice safety to obtain more food. The model predicted a general increase in vigilance with reserve levels, but the effect interacted with group size. In small groups, vigilance increased rapidly with the level of reserves before reaching a plateau. By contrast, in larger groups, vigilance was independent of state until reserves reached a high level. I discussed this effect in Chapter 4.

Build-up of reserves occurs through foraging. Therefore, the stable level of vigilance should vary with food intake rate. The model predicted a decrease in vigilance when individuals can obtain food at a higher rate. If food intake rate increases with food density, the model suggests that vigilance decreases with food density as long as time constraints operate, that is, as long as starvation is a real possibility.

5.3.3   Recent Developments

Here, I explore two recent developments that have challenged some of the traditional assumptions made in models of anti-predator vigilance.

5.3.3.1   Temporal Variation in Predation Risk

To document the group-size effect on vigilance, researchers need to measure vigilance in groups of different sizes. As explained earlier, it is possible to manipulate group size directly or to document vigilance opportunistically in different groups as they are encountered. Animals are expected to adjust their vigilance depending on the size of their current group. A recent model suggested that variability in group size across time could actually drives changes in vigilance (Lima and Bednekoff, 1999b).

In their model, Lima and Bednekoff applied the logic of the risk allocation hypothesis to vigilance in groups. The risk allocation hypothesis suggests that animals facing temporal changes in predation risk should allocate more anti-predator effort at times of greater risk. If animals expect to split their time between a small and a larger group, predation risk should vary temporally because individuals in large groups are more protected. Now, if an individual needs to accumulate a fixed amount of resources over the entire time period, how should it allocate anti-predator vigilance in these different groups to maximize fitness? The model showed that individuals should be more vigilant in the riskier situation, namely, when group size is smaller. Without variability in group size over the foraging time horizon, vigilance should be independent of group size.

The implication is that temporal fluctuation in group size can drive the group-size effect on vigilance. I am not aware of any direct test of this hypothesis with respect to anti-predator vigilance. An odd feature of the model is that group size is imposed on foragers for a fixed period of time. What prevents foragers from choosing the group size they like was not specified. In addition, the only source of safety in the model is provided by group size. However, foraging animals can seek shelter in a refuge like a bush or a hole in the ground. When a refuge from predation is available, animals in a large group could feed faster, at the expense of vigilance, to seek shelter rather than stay exposed during the whole period, questioning the prediction that vigilance is independent of group size if group size is constant (Beauchamp and Ruxton, 2011). This model challenges assumptions at the very core of traditional anti-predator vigilance models. Despite the previous caveats, more attention should be allocated to testing its predictions.

5.3.3.2   Negotiating the Outcome

This recent model challenges the mechanism responsible for achieving the stable solution in the vigilance game (Sirot, 2012). In a typical games model, each player makes a sealed bid, not indicating by any means their next move. In the schoolyard game of rock–paper–scissor, a move by a player is not motivated by any clue about what the other player is going to play next, but rather by the expectation that the other player is going to select a tactic that will maximize success (in this case, playing any of the three moves randomly and with equal frequencies). Essentially, if the players have the same ability and the same goal, it is actually irrelevant to know which move the opponent is going to select next.

In the vigilance game, any unilateral deviation from the evolutionarily stable strategy by any player reduces fitness. As we saw earlier, this evolutionary stable level of vigilance is typically much lower than the cooperative (but unstable) solution in which every group member adopts the level of vigilance that maximizes fitness. If a deviation from this optimal solution (i.e. cheating) cannot be prevented, then there is little value in watching companions to adjust vigilance levels since everyone should converge on the stable solution.

What would happen if bids were not sealed or, in other words, if individuals can actually negotiate the outcome of the game? This is the situation that Sirot modelled in his vigilance game. The model explores the tendency to initiate vigilance if others are not vigilant and the tendency to become vigilant if others are vigilant, making vigilance sensitive to the behaviour of companions in the group (in the actual model, the game involved only two players). When predators preferentially attack non-vigilant group members, the model predicts synchronization of vigilance amongst group members, which reduces the chances of being the only non-vigilant individual in the group during an attack. By contrast, a random striking strategy by the predator favours the opposite tendency: individuals tend to be vigilant when the others are not, and vice versa. This coordination of vigilance ensures a high probability of detection at the group level. These different vigilance scenarios illustrate how a strategy of responding to the vigilance of companions can lead to clearly observable patterns of vigilance at the group level. I review the empirical evidence for synchronization and coordination of vigilance in Chapter 7. Obviously, watching the level of vigilance maintained by others in the group can be costly in terms of time, but a recent model suggested that monitoring of only a few group members might be sufficient to generate discernible patterns of vigilance at the group level (Beauchamp et al., 2012). The recent finding that animals can solicit information from their neighbours to assess their vigilance also fits with the idea that bidding in the vigilance game might be an open one (Hare et al., 2014).

5.4 VALIDITY OF THE ASSUMPTIONS

All the models explored thus far make several assumptions about the vigilance process. Here, I want to assess the relevance of these assumptions. This is an essential step before examining their predictions.

5.4.1 Vigilance Varies as a Function of Group Size

All models of vigilance implicitly assume that variation in group size alone drives adjustments in vigilance. As I discuss in the Chapter 6, many factors other than group size can drive changes in vigilance. If these factors correlate with group size, it is more difficult to establish a causal link between group size and vigilance.

Two approaches have been used to determine the direct impact of group size on vigilance. The first obvious approach consists in manipulating group size. In an experiment, any variation in vigilance could be attributed to group size alone as long as other potentially confounding factors are carefully controlled. Manipulation of group size in the field is typically restricted to animals like invertebrates with a limited range of movements (Fordyce and Agrawal, 2001). Most experimental studies, therefore, are conducted in the laboratory. Laboratory studies showed that vigilance typically decreases with group size, supporting the expectation that group size alone can drive changes in vigilance (Beauchamp and Livoreil, 1997; Blumstein et al., 1999; Cézilly and Brun, 1989; Gosselin-Ildari and Koenig, 2012; Lazarus, 1979b; Rieucau et al., 2010; Vahl et al., 2005).

The second approach relates changes in vigilance to real-time changes in group size. Real-time changes in group size are so rapid that all potential confounds have little time to intervene. As an example of this approach, a recent study showed that visually-isolated group members immediately decreased their vigilance after hearing broadcasted calls simulating the presence of nearby companions (Radford and Ridley, 2007) (Fig. 5.6). Birds in flocks also altered their vigilance after the departure or the arrival of a group member (Beauchamp and McNeil, 2004; Popp, 1985; Roberts, 1995; Sullivan, 1984). While all of the previous studies support the expectation that group size can drive changes in vigilance, it is important to remember that group size might be but one of several factors contributing to the group-size effect on vigilance.

5.4.2 Trade-Off Between Vigilance and Foraging

If vigilance does not interfere with foraging, group size is not expected to exert any influence on vigilance. As shown in Chapter 3, vigilance and alternative activities are not necessarily mutually exclusive. Generally, the trade-off between foraging and vigilance will not be very costly when animals can overlap different processes during vigilance (Baker et al., 2011; Fortin et al., 2004a). While vigilance and foraging may not be as incompatible as once thought, few researchers would argue that vigilance is typically cost-free.

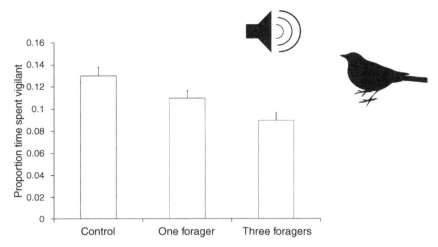

FIGURE 5.6   Real-time adjustments in vigilance. Visually isolated foraging pied babblers became less vigilant after hearing broadcasted contact calls from one or three companions as opposed to background noise (control). Means and standard error bars are shown. *(Adapted from Radford and Ridley (2007).)*

5.4.3   Randomness in Scanning

The models reviewed earlier assume that scans are initiated at random times. Recall that it is necessary to specify when scans are initiated to calculate the probability of detecting an approaching predator. Two types of randomness have been distinguished (Bednekoff and Lima, 1998a). Instantaneous randomness reflects the assumption that scans are initiated with the same probability regardless of how long an animal has been non-vigilant. Sequential randomness is a corollary of instantaneous randomness. With a random scan initiation process, the length of successive, or sequential, inter-scan intervals should be independent. Independence of scanning amongst group members, which is a further corollary of instantaneous randomness applied this time to all group members, will be covered in Chapter 7.

5.4.3.1   Instantaneous Randomness

The empirical distribution of inter-scan intervals should follow the negative exponential distribution if a Poisson process underlies scan initiation. The negative exponential distribution is an ever-decreasing function of inter-scan interval duration with no hump as in the normal distribution. Some early studies found a good fit between empirical distributions of inter-scan intervals and the negative exponential distribution (Bertram, 1980; Caraco, 1982; Studd et al., 1983), but the fit proved quite poor in others (Elcavage and Caraco, 1983; Pöysä, 1987; Sullivan, 1985). Typically, very short or very long intervals occurred less frequently than expected (Hart and Lendrem, 1984; Lendrem et al., 1986). The

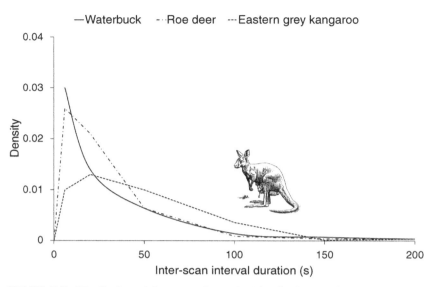

FIGURE 5.7 **Distribution of inter-scan intervals.** The distribution of inter-scan interval durations showed a hump in European roe deer (long dashed line), and Eastern grey kangaroo (short dashed line), but declined more rapidly in waterbuck (solid line). Estimated probability density functions are shown for each species. None of these distributions fitted the expected negative exponential distribution. *(Adapted from Pays et al. (2010).)*

distribution of inter-scan intervals actually contained a hump in several species (Beauchamp, 2006; Carro and Fernandez, 2009; Ross and Deeming, 1998; Sullivan, 1985). The negative exponential distribution also poorly fitted the distribution of inter-scan intervals in a recent study of three mammalian species (Pays et al., 2010). In this study, a humped distribution was apparent for the Eastern grey kangaroo and European roe deer, two species experiencing low predation risk. In the waterbuck, which faces a higher predation risk, the distribution of inter-scan intervals showed no hump but still poorly fitted the negative exponential distribution (Fig. 5.7).

The real question is not so much whether the empirical distribution of inter-scan intervals fits the negative exponential distribution, but rather why this particular pattern of scanning is expected in the first place. Recent theoretical work showed that a pattern of regular scanning, which would produce a more humped distribution of inter-scan intervals, is in fact superior to random scanning to detect sources of disturbances that occur randomly (Bednekoff and Lima, 2002). In fact, a random scan initiation pattern might only be relevant to prey animals that face observant predators (Bednekoff and Lima, 2002; Scannell et al., 2001). If prey animals showed a regular pattern of scanning, an observant predator could take advantage of predictable periods of low vigilance to launch attacks. By adopting a variable scanning pattern, prey animals instil uncertainty in the minds of observant predators. The first step before testing the fit of any model to an empirical distribution of inter-scan intervals should be to identify the type of predators faced by the study prey animal species.

5.4.3.2 Sequential Randomness

If scans are initiated randomly, there should be no predictability in the duration of successive inter-scan intervals. Any deviation from randomness will produce a correlation in the duration of successive inter-scan intervals. Spectral analysis provides a mathematical tool to uncover such a correlation by detecting periodicity in a long series of events (such as scans and inter-scan intervals) plotted as a function of time. Researchers used spectral analysis to detect periodicity in the rises and falls of inter-scan interval durations collected over time in different species (Desportes et al., 1990a; Desportes et al., 1990b; Desportes et al., 1989; Desportes et al., 1990c; Ferrière et al., 1996; Quenette and Desportes, 1992). However, claims that vigilance sequences in various animal species show significant periodicity have been controversial. A re-analysis of some of the original data revealed much fewer significant results (Suter and Forrest, 1994). Other researchers pointed out that a correlation between the duration of successive inter-scan intervals might be driven by independent responses to the same external disturbances, which would tend to produce temporal clusters of short inter-scan intervals (Ruxton and Roberts, 1999). With respect to the sequential randomness assumption, the evidence from spectral analysis is considered weak (Bednekoff and Lima, 1998a).

An alternative approach classifies inter-scan intervals as short or long, and seeks runs of such intervals in vigilance sequences, which would indicate that consecutive intervals tend to be similar in duration. Longer runs than expected by chance have been documented in flamingos (Beauchamp, 2006), greater rheas (Carro and Fernandez, 2009) and captive black tufted-ears marmosets (Barros et al., 2008), suggesting a certain level of predictability in successive inter-scan intervals. A more sophisticated statistical analysis of the sequential randomness hypothesis also showed considerable predictability in the duration of successive inter-scan intervals in other species (Pays et al., 2010).

Processes that drive non-random scanning patterns in animals are poorly known. I investigate two potential mechanisms in Chapter 7, namely, synchronization and coordination. The consequences of non-random scanning patterns for the group-size effect are not known.

5.4.4 Advantages to Detectors

What is the evidence regarding the assumption that direct detectors enjoy an advantage over non-detectors during an attack? An advantage to detectors was suspected in Thomson's gazelle groups because cheetahs appeared to target the less vigilant group members (FitzGibbon, 1989). Another line of research suggests that less vigilant foragers escape more slowly after an attack, probably increasing their chances of being targeted. Slower escape for non-vigilant

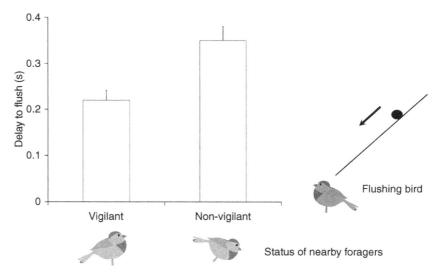

FIGURE 5.8   Advantage to detectors. Dark-eyed juncos flushed more quickly after a simulated attack on their neighbour if they were vigilant rather than non-vigilant at the time of the attack. A ball rolled down a chute simulated the attack that triggered the escape response in the targeted bird. Means and standard error bars are shown. *(Adapted from Lima (1994).)*

group members was documented in fish (Krause and Godin, 1996) and in birds (Elgar, 1986b; Hilton et al., 1999a; Lima, 1994). In a particularly clever experiment to address this issue, Lima (1994) used small balls rolled down a chute to simulate a predation attempt on a group of foraging birds. Birds targeted by the ball flushed to safety at if they were attacked by a real predator, but the ball could not be detected by other companions in the group (Lima, 1994). Lima noted how long vigilant and non-vigilant birds took to escape after a targeted bird flushed to safety. The results showed that vigilant birds escaped sooner than non-vigilant birds, suggesting that direct detection is indeed advantageous (Fig. 5.8). While rather convincing, the evidence overall is surprisingly slim for such a vital mechanism. Recall also that adopting a supposedly non-vigilant posture failed to affect escape time in blue tits (Kaby and Lind, 2003).

5.4.5   Improved Predator Detection in Groups

The many-eyes effect predicts that the detection of threats should occur more quickly in larger groups. Without this advantage, group members could only rely on dilution or other effects to reduce their risk of capture.

It is rather challenging to demonstrate this effect empirically. Predation attempts tend to be rare or difficult to witness. In observational studies, there is no control over the distance at which predators launch their attacks. Empirical studies thus typically rely on trained predators (Kenward, 1978) or on

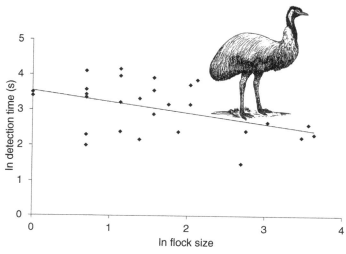

FIGURE 5.9   Detection distance and group size. Foraging emus detected an approaching person on foot more quickly in larger flocks. Notice the natural logarithm scale for the two axes. *(Adapted from Boland (2003).)*

simulated attacks using model predators (Powell, 1974) or approaching humans (Altmann, 1958). Even in carefully planned mock attacks, it might also be difficult to determine when detection actually takes place if prey animals do not show overt responses after detection or delay escape responses (Ydenberg and Dill, 1986).

In a classic study with a trained goshawk attacking from a standard distance, Kenward (1978) showed that as the number of prey wood pigeons increased in a group, the median detection distance also increased, as would be expected from the many-eyes effect. This pattern has also been reported in other species of birds (Boland, 2003; Cresswell, 1994) and mammals (FitzGibbon, 1990; Hoogland, 1981) (Fig. 5.9).

While it is possible to standardize predator behaviour to some extent in detection studies, the level of vigilance maintained by prey animals can rarely be controlled. Detection distance will increase if prey animals are more vigilant. Any factor that influences vigilance in a group will thus have an impact on detection ability. Consider the effect of group size on detection. Since each individual in a large group typically invests less time on vigilance, the level of vigilance maintained at the group level, which is the main factor affecting detection, will increase with group size but more slowly than expected on the basis of the vigilance maintained by a solitary forager. This means that the corporate ability to detect threats is unlikely to be near maximum in groups and will also vary with group size. Without an effective control on collective vigilance, detection distances might be difficult to compare across contexts.

5.5 CONCLUSIONS

Theoretical modelling has been instrumental in the development of vigilance research over the years. The model by Pulliam (1973), in particular, spearheaded subsequent empirical research on the group-size effect on vigilance and the many-eyes hypothesis. While it is easy with hindsight to highlight weaknesses in early vigilance models, there is no doubt as to their impact on the field of animal vigilance research. Recent models have addressed some of these issues, and added a wealth of new predictions awaiting empirical scrutiny.

Processes controlling the initiation of scans are poorly understood. It is now clear that we need to consider the attack tactics of predators to successfully predict the optimal pattern of scan initiation. Putting predators back in the vigilance picture will no doubt increase the complexity of future models of anti-predator vigilance. Ruxton and I considered, for instance, a game between vigilant prey and observant predators, and predicted that prey animals should decrease their vigilance since the beginning of a foraging bout while predators should typically attack early (Beauchamp and Ruxton, 2012a; Wheeler and Hik, 2014). Integrating this kind of arms races between predators and prey in the modelling framework appears a promising avenue for future developments. Similarly, the possibility that prey animals within the same group negotiate which level of vigilance is optimal opens up a new area of research (Sirot, 2012). This approach might help us understand how prey animals converge on the equilibrium solution. Open bidding seems to imply that collective vigilance might not be maintained at a constant or at a high level at all times. This surely must have an impact on the ability of the group to detect approaching predators.

Models of anti-predator vigilance have for the most part taken a molar approach, a term used in psychology to denote a methodology that focuses on outcomes rather than underlying processes. Vigilance models are thus mostly concerned with the overall allocation of time to vigilance, and typically fail to specify how long scans and inter-scan intervals should last. Given that time is limited, the best balance between foraging and vigilance will likely reflect how any investment in time pays off in terms of food intake or predator detection. Models thus far have made simplifying, and sometimes restrictive assumptions about how investment in time translates into food intake and predator detection.

Researchers have long suspected that the quality and quantity of information obtained during a scan vary with scan duration (Desportes et al., 1991a), but this has generally been ignored in vigilance models. Empirical research certainly suggested that scans are not always short and often variable in duration (Cowlishaw et al., 2004; Creel et al., 2008; Cresswell et al., 2003; Lawrence, 1985; Lendrem, 1983a; McVean and Haddlesey, 1980; Popp, 1988b). Longer scans probably allow individuals to view a wider area, but may provide redundant information if the environment changes slowly. Longer scans could reduce the amount of time spent in transition from feeding to scanning (Studd et al., 1983). Longer scans might be needed in situations in which information about predators is difficult to obtain. Too short a scan, by contrast, may miss crucial information about an impending threat. Others have also cited the utility of scanning, namely, how easily scanning can result in predator detection (Lendrem, 1983a; Scheel, 1993). When utility is low, short scans might be preferred since long scans would be wasteful. All these considerations challenge the notion that scans should be short and constant in duration.

The relationship between food intake and inter-scan interval duration is also probably more complicated than assumed in vigilance models. When food searching efficiency declines with the rate of interruptions, it is likely that feeding bouts with more scanning will provide less food per unit time spent foraging. This might be the case for foragers searching for prey that are difficult to detect (Carrascal and Moreno, 1992; Fritz et al., 2002). How vigilance interferes with the acquisition of food is poorly known in general.

A molecular approach, a methodology that emphasizes the kind of proximate considerations described previously, appears like a natural next step in the modelling of anti-predator vigilance. A molecular approach would be able to predict the frequency and duration of scans. A molecular approach is a good candidate to bridge the gap between proximate factors underlying vigilance, such as sensory ability, and the functional approach adopted thus far.

Chapter 6

Animal Vigilance and Group Size: Empirical Findings

6.1 INTRODUCTION

Living in groups offers much protection against predators (Beauchamp, 2014b). Even before the predator comes into sight, the formation of groups can already reduce the risk of predation. Predators, on average, need more time to find individuals concentrated in one spatial location rather than scattered across the habitat (Brock and Riffenburgh, 1960; Ioannou et al., 2011). Even if a group of N foragers were more conspicuous to the senses of a searching predator than a single forager, the group is unlikely to be N times more conspicuous (Inman and Krebs, 1987). For instance, individuals can hide behind one another so that the overall surface available for detection is much less than predicted from the size of the group. The actual site where a group aggregates might also offer more protection against predators, say, by being less accessible, than the average site occupied by a single forager. For all these reasons, the rate of encounter with predators is expected to be lower for individuals in a group, thus decreasing individual predation risk.

Should an encounter occur, I have shown earlier that prey animals in a group are more likely to detect an approaching predator early, giving each one a greater chance of escape. Prey animals in a group can also retaliate against the predator and offer considerable resistance (Mukherjee and Heithaus, 2013). Should a chase occur, the fleeing prey animals can confuse the predator, which will reduce the probability of individual capture (Ioannou et al., 2009). When evading predators, some individuals occupy more sheltered position in the group, again reducing the risk of capture for some if not all group members (Hamilton, 1971). All these effects strongly suggest that predation risk is likely to decrease as the size of a group increases. From the point of view of anti-predator vigilance, reduced predation risk should translate into a decrease in vigilance in larger groups.

This prediction was anticipated by early researchers like Galton (1871) and Lack (1954), and formalized mathematically for the first time by Pulliam (1973). In the previous chapter, I explored why vigilance is expected to decrease with group size and reviewed key assumptions. Here, I examine the empirical evidence for this effect, which has fostered one of the most active research

Animal Vigilance. http://dx.doi.org/10.1016/B978-0-12-801983-2.00006-1

141

programmes in animal behaviour over the last four decades. I will explore the magnitude of the group-size effect on vigilance, review alternative explanations for this effect, and discuss how some ecological factors like sex and food density shape and can potentially confound the expected relationship.

Few researchers have tested quantitative predictions of anti-predator vigilance models. In a quantitative test, observed levels of vigilance in groups of different sizes are pitted against predicted values. Such tests require detailed knowledge about how quickly a predator can approach the group, the distance between predator and prey, and the length and frequency of scans, amongst other factors (Elgar and Catterall, 1982; Pulliam et al., 1982). Given that such information is difficult to get, researchers tend to focus instead on the qualitative prediction that vigilance decreases with group size.

6.2 META-ANALYSIS

The first study devoted to the qualitative effect of group size on vigilance was published nearly 40 years ago (Dimond and Lazarus, 1974). In European white-fronted geese foraging in agricultural fields during the winter, the percentage of time individuals spent with the head up, which is a measure of vigilance, declined sharply with flock size. At the same time, the total number of individuals with the head up in the flock increased with flock size (Fig. 6.1). This early result provided support for the group-size effect on vigilance, and also suggested that

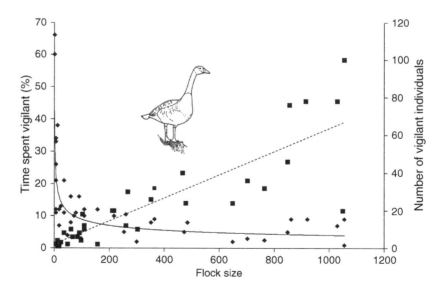

FIGURE 6.1 Vigilance and flock size. In European white-fronted geese, the amount of time each individual allocated to vigilance (time spent with the head up in this case, black diamonds) decreased in larger flocks. The total number of individuals with the head up (black squares); however, increased with flock size. Trend lines are shown for each plot. *(Adapted from Lazarus (1978)).*

despite the reduction in individual vigilance, the level of vigilance maintained at the group level was never lower than that of a solitary forager and was, in fact, much higher in larger groups. Geese in flocks reduced their investment in anti-predator vigilance at no increased risk to themselves.

Hundreds of studies subsequently investigated the group-size effect on vigilance in birds and mammals. Early reviews of this effect found overwhelming support for the prediction (Elgar, 1989; Quenette, 1990). However, several studies showed no effect of group size on vigilance or even the opposite effect. In these early reviews, support for the prediction was based on the p-value of the statistical relationship between vigilance and group size. If the p-value was statistically significant, the study supported the prediction, but not otherwise.

The p-value of a statistical test is notoriously unreliable when evaluating the evidence for a prediction. Tests conducted with a small sample size are certainly less likely to yield statistically significant results. Statistically significant results are also less likely for weaker relationships. Even when a result is statistically significant, the p-value tells us little about the magnitude of the effect. In the end, we want to be able to determine how much of the variation in vigilance can be explained by group size. As a concrete example of the pitfalls of using p-values, vigilance in golden marmots was significantly associated with group size, but group size was a rather poor predictor of vigilance (Blumstein, 1996).

Standardized estimates of effect sizes are best to characterize the relationship between vigilance and group size in a large number of studies. Meta-analysis has become the tool of choice to compare estimates of effect sizes across studies. Briefly, meta-analysis pools results from a number of studies that are all testing the same hypothesis with the aim of providing a quantitative estimate of the evidence for that hypothesis (Lipsey and Wilson, 2001; Nakagawa and Santos, 2012). Estimates from different studies are typically weighted using variables such as sample size that are thought to improve the reliability of the findings. The coefficient of correlation between vigilance and group size provides a standardized estimate easy to calculate for a meta-analysis of vigilance.

Meta-analysis provides a handle for the control of type II error rate; the probability of failing to reject a false null hypothesis. A statistically borderline result can cast doubts on the validity of the hypothesis in one study. However, our confidence in the hypothesis increases when many borderline cases are reported along with other less equivocal results.

I conducted a meta-analysis to examine the magnitude of the group-size effect on vigilance in birds (Beauchamp, 2008b). To this end, I calculated the correlation coefficient between vigilance and group size in each available study, and pooled all estimates to provide an overall assessment of the hypothesis that vigilance decreases with group size (Fig. 6.2). The meta-analysis revealed that, overall, time spent vigilant by an individual decreases significantly with group size. The magnitude of the effect, as measured by the correlation coefficient, varied widely amongst studies. In addition, many confidence intervals around

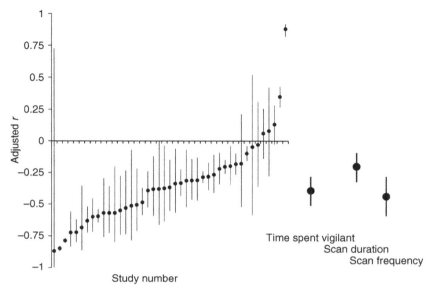

FIGURE 6.2 **Meta-analysis of vigilance in birds.** Studies typically reported a negative correlation between time spent vigilant and group size ($n = 43$). The correlation coefficient is shown as a dot for each study and the bar covers the 95% confidence interval around each correlation coefficient. The relationship between vigilance and group size is not statistically significant when the confidence interval includes the value of 0. The overall effect of group size on time spent vigilant, scan frequency, and scan duration is shown on the right-hand side. *(Adapted from Beauchamp (2008b)).*

correlation coefficients included the value of 0, which means that the relationship between vigilance and group size was not always statistically significant.

The overall magnitude of the effect was estimated at −0.4 for time spent vigilant, a medium effect. This value is considerably higher than many effect sizes reported in the ecology and evolution literature (Møller and Jennions, 2002). The overall magnitude varied depending on the proxy used to measure vigilance. The magnitude of the group-size effect was similar for scan frequency and time spent vigilant, but smaller for scan duration (Fig. 6.2). Recall that earlier models of anti-predator vigilance typically assumed that scans are short and fixed in duration. Therefore, a correlation between scan duration and group size was not expected, which suggests that this assumption should be relaxed in future models. Nevertheless, the stronger correlation with scan frequency implies that adjustments in vigilance are more likely to involve changes in scan frequency than scan duration.

I want to discuss two potential weaknesses of the meta-analysis of vigilance. First, a meta-analysis necessarily depends on the studies available for analysis. However, it might be the case that studies reporting a statistically significant outcome are more likely to be published, leading to the so-called file-drawer effect. This type of publication bias can be qualitatively assessed with a funnel plot. In a funnel plot, standardized estimates, such as the correlation coefficient that I used in the meta-analysis of vigilance, are plotted against a measure of the validity of a study, typically sample size. If the available studies represent an unbiased sample of all performed studies, the plot should show a funnel shape with a wider scatter at lower sample sizes. As sample size increases, the range of values should become narrower and eventually converge on the true effect size. Any part missing from the funnel shape suggests publication bias. In the meta-analysis of vigilance in birds, heterogeneity in the magnitude of the effect amongst studies was not related to sample size, suggesting that publication bias was probably not an issue here.

Second, different studies culled for a meta-analysis are assumed to provide independent estimates of the effect size. This is not likely to be the case for the meta-analysis of vigilance in birds for several reasons. One problem is that researchers tend to focus on a handful of species that are readily available and easy to study like the house sparrow or the European starling. This tendency, I believe, will underestimate the variability expected from truly independent studies. Another problem is that many of the species used in vigilance studies are closely related, making them likely to respond the same way to the same ecological challenges. Therefore, it would not be surprising if the correlation between vigilance and group size clustered around similar values for related species. These two sources of non-independence probably biased the confidence interval around the estimated effect sizes. A recent study showed that effect sizes can increase or decrease after controlling for relatedness, but in many cases the overall effect was no longer statistically significant (Chamberlain et al., 2012). Recent advances in meta-analysis provide tools to deal with lack of independence (Lajeunesse, 2009), and this issue should be addressed in future meta-analyses of vigilance.

6.3   WHY VIGILANCE FAILS TO DECREASE WITH GROUP SIZE?

The meta-analysis in birds revealed an overall medium effect of group size on vigilance. Nevertheless, many studies failed to document the predicted relationship. What are the reasons behind this failure? In the following, I discuss three possibilities.

6.3.1   Low Statistical Power

Statistically, studies with a small sample size might simply lack the power to detect a medium effect. In birds, at least, the magnitude of the effect was not related to sample size, which means that the failure to document the group-size effect on vigilance was equally likely in studies with small or large sample sizes. In the following, therefore, I focus on biological explanations for the lack of effect of group size. In general, given the expected medium effect of group size on vigilance, all but the smallest studies should be able to reject the null hypothesis.

6.3.2 Alternative Targets of Vigilance

The logic behind the decrease in vigilance with group size hinges on perceived predation risk. However, sources of vigilance not involving predators are certainly possible. As we saw in earlier chapters, individuals can use vigilance to monitor threatening conspecifics. Here, I argue that the need to monitor conspecifics should increase with group size and could interfere with the effect of group size on anti-predator vigilance for at least two reasons.

First, if individuals monitor each other to discover food sources, monitoring conspecifics should be more frequent in larger groups with more scrounging opportunities (Vickery et al., 1991). The expected decrease in anti-predator vigilance with group size might be compensated by an increase in vigilance aimed at companions to locate food sources (Beauchamp, 2001), resulting in little or no changes in overall vigilance with group size.

Second, conflict is likely to increase with the frequency of interactions amongst group members. The need to monitor threats within the group should thus increase in larger groups with more interaction opportunities. Increased aggression in larger groups, and the concomitant increase in monitoring effort, has often been invoked to explain why vigilance fails to decrease in larger groups (Barbosa, 2002; Blumstein and Daniel, 2002; Cameron and Du Toit, 2005; Teichroeb and Sicotte, 2012). Distinguishing between the targets of vigilance is necessary to determine whether vigilance aimed at conspecifics interferes with the group-size effect on anti-predator vigilance.

6.4 WHY THE MAGNITUDE OF THE GROUP-SIZE EFFECT VARIES?

In birds, the magnitude of the group-size effect on vigilance varied substantially amongst species. The magnitude of the effect can also vary within a species. For instance, the group-size effect on vigilance can be more pronounced for individuals of one sex or when predation risk is higher. I explore these patterns below. In general, published relationships between vigilance and group size show a funnel shape indicative of considerable variation within species in the magnitude of the group-size effect on vigilance (Box 6.1).

BOX 6.1 Evidence for Heterogeneity in the Group-Size Effect on Vigilance

When investigating the group-size effect on vigilance, researchers estimate the relationship between vigilance and group size by fitting a straight line to the scatter of data. This is typically done with a simple linear regression. An underlying assumption in a regression model is homoscedasticity (Sokal and Rohlf, 1995). Indeed, the dispersion of the vigilance measurements about the mean should be the same at all group sizes. A funnel shape, by contrast, is suggestive of heteroscedasticity because the data are more spread out at one end than the other. A funnel-shaped scatter often implies that variables other than the independent variable considered in the analysis influence the dependent variable (Rosenbaum, 1995). Such unmeasured variables interact with the independent variable and increase the scatter in the relationship.

Interactive effects are common in ecology in the form of limiting factors that prevent organisms from always showing a full response (Cade et al., 1999). As an example, consider the relationship between the number of trout in a stream and the ratio between the width and the depth of the stream, an index of habitat suitability (Dunham et al., 2002). If the ratio alone explained the number of trout in a stream, there should be little variation in the observed number of trout in streams characterized by a given ratio. However, trout number might also depend on other unmeasured factors such as temperature. The relationship between trout number and the stream ratio may be strong at low temperatures, but absent otherwise. When plotting the number of trout as a function of the stream ratio over the full range of temperature, the scatter in the data will appear funnel-shaped because the strong and the weak relationships are superimposed.

The occurrence of heteroscedasticity also implies that the magnitude of the relationship between any two variables may be poorly estimated. The standard linear regression only looks at the effect of the independent variable on changes in the mean of the dependent variable. The relationship estimated at the mean may not be representative of the real effect at work (Cade and Noon, 2003). Coming back to the fish example, the slope of the relationship between trout number and the stream ratio may become stronger and closer to the real value when focusing on higher percentiles of the distribution of trout number.

Quantile regression analysis can be used to assess heterogeneity in a scatter plot. The quantile regression extends predictions not only to the mean, but also to any percentile of the distribution (Koenker, 2005). One strategy is to run a regression for percentiles 0.2 and 0.8, which lie at the bottom and the top of the distribution, respectively. When the slopes at these two percentiles are statistically different, we have evidence for heteroscedasticity.

I analyzed 45 published scatter plots of the relationship between vigilance and group size in birds and mammals (Beauchamp, 2013b). Nearly 40% of the cases showed a statistically significant heteroscedasticity. In such cases, the slope from a standard linear regression was on average 59% smaller than that from the 0.8 percentile regression (Fig. 6.3).

Interaction with unmeasured variables is probably common in observational studies of vigilance. The quantile regression analysis also suggests that the magnitude of the group-size effect might be underestimated by quite a margin. Future studies can help us understand which ecological variables interact with group size to produce the wide scatter in the relationship between vigilance and group size.

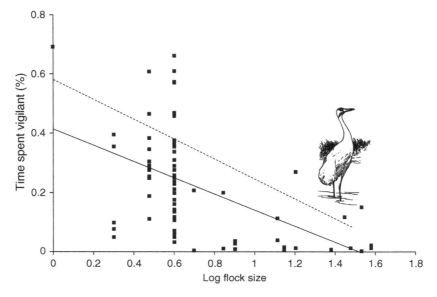

FIGURE 6.3 **Variability in the relationship between vigilance and group size.** In red-crowned cranes, time spent vigilant decreased with flock size, but the scatter in the data was more pronounced in smaller flocks. The quantile regression line estimated at the 80th percentile (dashed line) was significantly steeper than the standard linear regression line estimated at the mean (solid line), indicating the magnitude of the group-size effect on vigilance was not homogeneous. *(Adapted from Beauchamp (2013b)).*

6.4.1   Interaction with Sex

Differences in the magnitude of the group-size effect on vigilance in relation to sex have been noted in many studies in birds and mammals (Cameron and Du Toit, 2005; Childress and Lung, 2003; Li et al., 2012; Rieucau et al., 2012), but no clear pattern has emerged as to which sex shows the largest magnitude.

Sexual differences in the magnitude of the group-size effect suggest that the target of vigilance probably differs between males and females. In elk, for example, vigilance decreased with group size in females without calves, but not in females with calves or in males (Childress and Lung, 2003; Lung and Childress, 2007). Females with calves and males probably devoted part of their vigilance to monitoring threats within the group, which, as I argued earlier, can weaken the predicted decrease in anti-predator vigilance with group size. Ungulates in other species also devote much vigilance to monitoring rivals (Cameron and Du Toit, 2005; Li et al., 2012), which limits the scope of adjustments in anti-predator vigilance with group size. From a practical point of view, sexual differences in the magnitude of the group-size effect highlight the need to sample the sexes separately to document vigilance patterns.

6.4.2   Interaction with Predation Risk

Predation risk is also thought to interact with group size. Gradients in predation risk have been linked to several ecological variables, including distance to cover, position in the group, and predator density. The occurrence of such gradients allows us to examine how vigilance varies with group size when predation risk is high or low. Various patterns of interaction have been documented thus far, with the largest decrease in vigilance with group size occurring when predation risk is high (Frid, 1997; Lendrem, 1984; Lima, 1987a) or low (Lima et al., 1999; Manor and Saltz, 2003; Martella et al., 1995).

Why should there be an interaction between predation risk and group size? The effect of group size on vigilance was predicted to be more pronounced when predation risk is high (Frid, 1997) or low (Manor and Saltz, 2003). But without modelling the costs and benefits of vigilance as a function of group size at different levels of predation risk, it is not clear which way the predictions should go.

Extending earlier analyses (Bohlin and Johnsson, 2004; Grand and Dill, 1999), Bednekoff and Lima (2004) used just such a cost-benefit analysis to examine the allocation of time to vigilance in groups of different sizes under different levels of predation risk. Following the tradition of earlier vigilance models (Chapter 5), an increase in foraging effort (at the expense of vigilance) reduces the ability to survive attacks but increases the probability of avoiding starvation. The model showed that the magnitude of the group-size effect on vigilance should be larger when predation risk is low rather than high (Bednekoff and Lima, 2004) (Fig. 6.4). In other words, adjustments in vigilance in relation to predation risk should be more extensive in larger than in smaller groups. This is consistent with the pattern of change reported by Manor and Saltz (2003) but not with the one reported by Frid (1997).

An interaction between group size and sex or predation risk certainly illustrates how vigilance can vary with group size to different extent within the same species. The funnel-shaped scatter in many published relationships between vigilance and group size is also indicative of interactive effects. The next step is to identify which ecological factors can lead to such scatter in different species.

6.5   ALTERNATIVE HYPOTHESES TO EXPLAIN THE GROUP-SIZE EFFECT ON VIGILANCE

Traditional models of anti-predator vigilance are based on a trade-off between safety and foraging efficiency, and all predict a decrease in vigilance with group size. However, alternative hypotheses, based on different mechanisms, have been proposed to explain the effect of group size on vigilance. It is highly recommended to examine the relevance of these alternative explanations in any discussion of vigilance in groups.

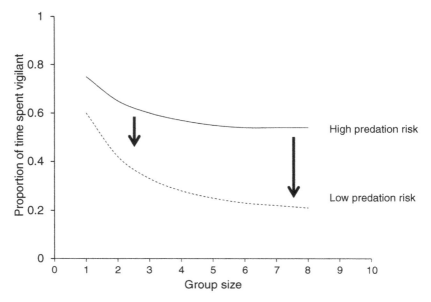

FIGURE 6.4 Interaction between group size and predation risk. Vigilance decreases with group size more extensively when predation risk is low rather than high, showing how predation risk modulates the effect of group size on vigilance. *(Adapted from Bednekoff and Lima (2004)).*

6.5.1 Edge Effect

As we saw in Chapter 4, spatial position in a group represents a major driver of vigilance in animals. Animals at the edges of a group are typically more vigilant than those in the middle. This simple fact can explain why vigilance decreases with group size even when group size has no effect on individual vigilance (Section 4.3.1). To reject this argument, it is necessary to show that vigilance decreases with group size at the edges and in the middle of the group.

6.5.2 Target of Vigilance

Unsuccessful foragers could use vigilance to obtain clues about the location of better food patches (Krebs, 1974). Because large groups tend to aggregate in rich food patches, the need to gather food-related information through vigilance should decrease with group size. This hypothesis predicts that vigilance precedes movements to a food patch. In addition, foragers should typically experience an increase in foraging success following relocation. This explanation has been largely ignored (Coolen and Giraldeau, 2003; Robinette and Ha, 2001), but deserves more attention especially in species exploiting food patches together.

6.5.3   Food Competition

Food competition represents the alternative explanation for the group-size effect on vigilance that has received the most attention. Vigilance models are usually concerned with predator detection and rarely consider the actual foraging process. Nevertheless, individuals must obtain food to survive, which means that vigilance should be sensitive to the amount and distribution of resources (Chapter 4).

When individuals compete for limited resources, a unilateral decrease in vigilance would allow one group member to obtain a disproportionate share of the resources and increase its foraging success relative to the others (Clark and Mangel, 1986). Faced with a unilateral defection, other group members should also decrease their own vigilance to exploit resources just as quickly, a process that will eventually bring vigilance down in the whole group. This is known as the milk-shake effect. Just how low vigilance can get in a scramble for limited resources depends on many factors. Foraging more quickly probably increases predation risk by shifting attention from threats to competitors. Exploiting resources more quickly also probably increases the energy cost of foraging (Shaw et al., 1995). Vigilance in a scramble for resources is expected to reach a new equilibrium value in which foraging gains from a decrease in vigilance just compensate loss of safety and rising exploitation costs (Beauchamp and Ruxton, 2003; Bednekoff and Lima, 2004; Lima et al., 1999). As competition increases in larger groups, the decrease in vigilance with group size constitutes an adaptive response to competition for limited resources.

How can we determine whether food competition of this kind influences vigilance? The model by Bednekoff and Lima (2004), which I described earlier, makes useful predictions. We saw earlier that without competition vigilance is expected to decrease with group size more strongly when predation risk is lower, yielding an interaction between group size and predation risk. It turns out that the magnitude of the group-size effect on vigilance, when only competition matters, is remarkably similar under low or high predation risk, yielding this time no interaction between group size and predation risk. When group size influences both competition and the probability of surviving an attack, the interaction between group size and predation risk remains in the same direction but more extreme. These various interaction patterns between group size and predation risk allow us to determine whether the group-size effect on vigilance is mostly driven by competition, by predation, or both (Fig. 6.5).

Another way of showing an effect of competition on vigilance patterns is to document vigilance when competition for food can be effectively ruled out, say, by providing an effectively limitless amount of food (Lima et al., 1999). However, while animals may not compete directly to obtain more food in this case, they may compete to leave the food patch as quickly as possible for the safety of a refuge (Beauchamp and Ruxton, 2007), which means that some competition can still persist.

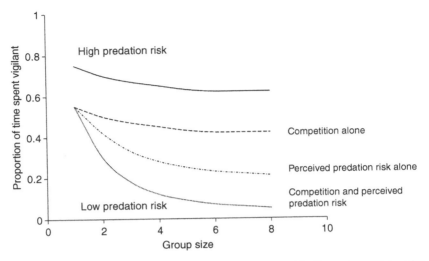

FIGURE 6.5 Interaction between group size and predation risk. The pattern of interaction between group size and predation risk can tell us whether food competition alone, predation risk alone, or their combination drives changes in vigilance. *(Adapted from Bednekoff and Lima (2004)).*

It is perhaps better to discount competition by actually manipulating the level of competition in the group. Such manipulations have been performed in a handful of species. In captive Tammar wallabies, the number of food bins available to the group was reduced to induce competition, but vigilance, unfortunately, did not decrease as predicted (Blumstein et al., 2002). In coots, competition was induced by providing a highly valued but scarce food (Randler, 2005a). Controlling for group size, coots maintained a lower vigilance when eating the scarce food, supporting the competition hypothesis (Fig. 6.6). In an elaborate laboratory experiment involving video screens, nutmeg mannikins decreased their vigilance to a greater extent when watching virtual companions feeding more intensely at the expense of vigilance (Rieucau and Giraldeau, 2009). One difficulty here is that foragers witnessing less vigilance might have perceived the situation as less dangerous, leading to a decrease in vigilance. In addition, since virtual companions exerted no effect on the food supply available to foragers, it remains difficult to ascribe the changes in vigilance to real competition.

How competition for resources shapes vigilance patterns has largely been ignored in the literature. Nevertheless, the extensive review on the drivers of vigilance (Chapter 4) made it abundantly clear that competition for food is certainly not the only factor influencing vigilance. Competition probably plays a limited role in animals that feed on plentiful resources or in species that can overlap foraging and vigilance. More empirical evidence is needed to determine to which extent vigilance is adjusted to competition intensity.

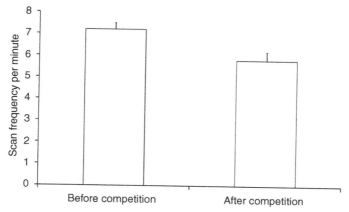

FIGURE 6.6   Food competition and vigilance. Coots feeding on a scarce but valuable food re-source scanned less often than when they fed earlier on an abundant food source. Sacrifices in safety can increase foraging speed when animals compete for limited resources. Mean and standard error bars are shown. *(Adapted from Randler (2005a)).*

6.6   CONCLUSIONS

Empirical research on the effect of group size on animal vigilance has cast a wide net; it has included a broad range of avian and mammalian species and combined observational as well as experimental studies. This vast literature has now been subjected to a meta-analysis in birds, which has shed some light on the magnitude of the group-size effect and its variation both within and amongst species. Extending this approach to mammals represents a next step sure to help us validate earlier conclusions based on avian studies.

Research on the group-size effect on vigilance has been mostly qualitative. The next obvious step is to test quantitative predictions from vigilance models. Models of vigilance predict cooperative and selfish solutions, and it would be interesting to see which solution tends to be adopted in different study systems. In addition, testing quantitative predictions will provide much needed feedback for the way we model mechanisms responsible for the decline in vigilance with group size, including collective detection and risk dilution.

Research on the group-size effect addresses ultimate questions about vigilance by testing predictions from adaptive models. Proximate mechanisms and developmental issues, by contrast, have received little attention. For instance, we still have very limited information about how age influences the relationship between vigilance and group size. It may turn out that animals have to learn that large groups are safer. However, few studies have focused on the development of vigilance (Loughry, 1992; Sullivan, 1988) (Chapter 3). The perception of group size is also poorly known despite playing a central role in the group-size effect on vigilance (Bekoff, 1996; Elgar et al., 1984). Do animals count group members or approximate group size with indirect indices like the area covered by the group or the noise generated by group members? Studies on this topic may reveal constraints on the ability to adjust vigilance to group size. An inability to accurately estimate group size in larger groups might explain why vigilance decreases more slowly as group size increases.

To document the relationship between vigilance and group size, researchers must have an operational definition of group size. However, where to draw the line that circumscribes a group remains a problematic issue. The most common method is to simply count the number of individuals within a fixed radius from the centre of the group (Stankowich, 2003). In many cases, the maximum distance that allows communication between group members is not known, which means that the so-called radius of interaction often has no real biological value. Many studies have shown that nearby companions exert more influence on vigilance (Elgar et al., 1984; Fernández-Juricic and Kowalski, 2011; Lima and Zollner, 1996), and it is also clear that this limit varies from species to species depending on their perceptual abilities. Our knowledge of sensory ecology is relatively scant, which makes it difficult to assess the relevance of our assumptions about group membership (Fernández-Juricic, 2012).

Overall, a broader approach to the study of vigilance in groups, including both proximate and ultimate questions (Hofmann et al., 2014), and an emphasis on testing quantitative predictions are likely to keep vigilance at the forefront of active research in animal behaviour.

Chapter 7

Synchronization and Coordination of Animal Vigilance

7.1 INTRODUCTION

Models of animal vigilance make simplifying assumptions about scanning behaviour. In Chapter 5, we saw, in particular, that scan initiation is thought to be governed by a memoryless process. With such a process, the probability of initiating a scan stays the same regardless of the amount of time that has elapsed since the last scan. As a corollary, the length of successive intervals between scans also becomes independent. Substantial deviation from the two types of randomness has been documented in many species, casting doubts about the validity of these assumptions. In this chapter, I focus on yet another assumption about scanning behaviour: the notion that the initiation of a scan is independent amongst group members.

The assumption of independent scanning is fundamental; without it, it is impossible to assume that each group member contributes equally to predator detection. With full independence, it becomes much easier, mathematically speaking, to calculate the probability of not detecting an approaching predator at the group level; we can simply raise to the power of N (the size of the group) the probability that any group member fails to detect the approaching predator.

Independent scanning need not imply that group members are oblivious to one another. Monitoring conspecifics is necessary for various reasons. Potentially threatening individuals certainly require some monitoring. When adjusting vigilance, animals must also monitor companions to estimate group size. For collective detection to work, individuals must also spend time monitoring neighbours for signs of alarm. However, in theory, there should be no need to determine how many neighbours are vigilant or how vigilant they are. When modelled as a game between group members, vigilance is predicted to reach an evolutionary stable level from which any unilateral deviation is costly. The cost of deviating from the stable level of vigilance implies that all group members should adopt the same level of vigilance. Therefore, there is no need to monitor the actual vigilance maintained by others.

The previous discussion suggests that at equilibrium individuals should initiate scans independently from one another. The actual timing of a vigilance bout will be

Animal Vigilance. http://dx.doi.org/10.1016/B978-0-12-801983-2.00007-3

155

dictated solely by the realization of whichever process governs scan initiation. Due to the random nature of scan initiation, the temporal sequence of scans and inter-scan intervals will differ amongst individuals but fluctuate about the same mean.

Consider an analogy with slot machines in a casino. When many patrons operate slot machines simultaneously, there will always be times where several win at the same time, but also times where no one wins. When many patrons conspicuously win at the same time, a slot machine operator in a losing streak perceives the odds of winning from any terminal as much higher than they really are, a property exploited by casino owners to entice customers to play longer. A savvy gambler, like a prey animal in a group, should not monitor the successes and failures of neighbours because all slot machines are the same and each one operates independently of the others.

When viewed as the outcome of a stochastic process, the number of vigilant individuals at any one time is expected to fluctuate widely in the group. By chance many foragers or none at all may be vigilant at the same time. Oddly enough, independent scanning is likely to lead to transitory periods where too much or too little safety prevails in the group.

When faced with the task of detecting a threat, independent monitoring by group members is definitely not the tactic that we as humans would choose. A sentinel system appears much more suited to the task of reliably detecting a threat. In a sentinel system, individuals take turn monitoring the surroundings. This tactic ensures a more even vigilance effort at the group level, and also allows maximum rest for each group member. To coordinate vigilance effectively in a sentinel system, it is imperative to monitor the vigilance of others to ensure constant surveillance. This is one instance where the assumption of independent scanning is clearly violated. Later in this chapter, I will explore factors that promote coordination of vigilance in prey animal groups.

If coordination of vigilance reduces the number of times during which no one is vigilant, synchronization of vigilance produces the exact opposite effect. Synchronization of vigilance implies that a prey animal is more likely to be vigilant when others are vigilant, and vice versa. Synchronization is thus expected to produce transitory periods during which too many or too few group members are vigilant at the same time. Synchronization of vigilance, like coordination, violates the assumption of independent scanning in a group. Recent models have explored how such a counterintuitive tactic can actually increase individual fitness in groups. In this chapter, I first examine the adaptive value of vigilance synchronization and then review the available evidence.

7.2 SYNCHRONIZATION OF VIGILANCE

Two bodies of theories can explain why individuals would copy the level of vigilance maintained by their neighbours. The first theory emphasizes the estimation of predation risk while the second theory focuses on attack tactics by predators.

7.2.1 Vigilance and the Estimation of Predation Risk

Solitary animals can only rely on their own effort to gather information about predators. By contrast, individuals in groups can complement their own effort with information gleaned from the behaviour of other group members. In technical terms, animals in groups have access to personal as well as public information (Danchin et al., 2004). This logic can be applied to the estimation of predation risk. Vigilance allows individuals to gather personal information about potential threats. In addition to this personal source of information, group members could also use public information generated by the vigilance behaviour of others (Rands, 2010).

Thus far, I have presented vigilance as a means to detect threats. Viewed this way, there is no information available from the actual scanning behaviour of neighbours. Indeed, with collective detection, group members need only pay attention to the outcome of vigilance, namely, whether or not a vigilant neighbour shows signs of alarm. Vigilance, however, can also be viewed as a means to allay fear. The notion that vigilance betrays fear is an old one. For example, increased feeding when chicks are in groups rather than alone was attributed to a reduction of fear (Tolman, 1965). Fear of attacks by predators was also thought to lead to uneasiness in wood pigeons, which again would be betrayed by greater vigilance (Murton et al., 1971). Vigilance can thus reflect fear and indirectly the risk of predation.

In the framework of fear, a prey animal initiates a scan because it believes the situation has become more dangerous (Sirot and Pays, 2011). Therefore, the level of vigilance maintained by neighbours can act as an indirect cue of potential danger. Fear should increase when more neighbours are vigilant. By contrast, relaxed neighbours suggest a low predation risk. In response to this public information about predation risk, a forager may choose to increase or decrease its own vigilance. Paying attention to the vigilance of others makes sense in the context of gathering public information about predation risk.

Viewed as public information, vigilance can become contagious; it can spread or not within a group because individuals are influenced by the behaviour of other group members. How do we incorporate public information in an anti-predator vigilance model? In McNamara and Houston's 1992 model, the probability of death upon attack for a given individual in the group was contingent on three distinct possibilities: (1) the focal individual detected the predator on its own, (2) the focal individual failed to detect the predator but at least one other group member did, and (3) no individual in the group detected the predator. The probability that each of these events would occur was then multiplied by the probability that the predator would target the focal individual out of the N individuals available in the group. This product gives us the overall probability of capture or death. The probability of death can be thought of as the actual risk of predation, and the model clearly showed that vigilance is expected to increase when predation risk increases.

Sirot (2006) transformed the actual predation risk into a perceived predation risk, the risk estimated by an individual after gathering its own information and watching the behaviour of other group members (Sirot, 2006). Sirot considered that the predation risk from McNamara and Houston's model should be multiplied by a factor that combines the personal as well as the public information available to a forager about predation risk. In the simplest case where no public information is used, the perceived predation risk represents some baseline estimate for the habitat (i.e. an estimate known from previous experience or encoded in the genes) adjusted to the situation experienced by the focal animal. The current situation involves factors such as group size and the relative efficiency of collective detection, and is calculated using the actual predation risk. Now, if public information is used, foragers weigh the aforementioned personal estimate of predation risk with a social information index. When this index takes on a small value, the perceived predation risk reflects mostly personal information, but as the value of the index increases, the balance shifts to public information. This public information is based on the proportion of vigilant neighbours, a sign of the prevailing anxiety in the group.

The model showed that when public information is ignored, the evolutionary stable level of vigilance decreased as proportionately more neighbours are vigilant because foragers can use the extra vigilance in the group to reduce their own (Fig. 7.1). However, when public information influences the estimate of predation

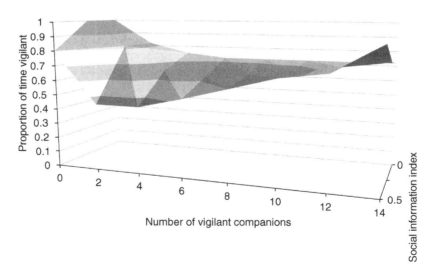

FIGURE 7.1 Contagion model of vigilance. When more individuals in the group are vigilant, the optimal level of vigilance of a forager is predicted to increase if individuals take into account the vigilance behaviour of others (large values of the social information index). Vigilance, however, is not expected to increase when social information is ignored (small values of the social information index). The occurrence of many vigilant companions in the group can betray an imminent threat in which case the best response is to increase one's own vigilance. *(Adapted from Sirot (2006)).*

risk, vigilance can actually increase when more neighbours are vigilant. The occurrence of many vigilant neighbours can betray a greater perceived risk of predation, and the best response by a forager is to increase its own vigilance.

The use of public information about predation risk can produce synchronized vigilance. In such a group, periods of extreme vigilance will be interspersed with periods of little vigilance. Such temporal waves of vigilance, unfortunately, are also predicted by another mechanism involving predator targeting behaviour (see section 7.2.2), and cannot be used to support the public information hypothesis.

Is there any evidence for contagion of fear or vigilance in animals? Some evidence comes from a recent laboratory experiment in voles (Eilam et al., 2011). Voles are naturally anxious animals, but individuals vary quite extensively in their level of anxiety. The level of anxiety can be established by measuring the amount of time individuals are willing to spend alone in the open area of a maze. It is generally agreed that animals willing to stay longer in the open area, potentially exposed to more danger, are less anxious. After establishing the baseline level of anxiety in many voles, researchers put the voles together in a cage and exposed them to repeated (but not life-threatening) attacks by one of their natural predators, the barn owl. Researchers then documented the level of anxiety of each individual separately after the attacks using the same methodology as before. Not surprisingly, all individuals tended to spend less time in the exposed area (Fig. 7.2). The unexpected finding was that the naturally less anxious voles converged on the level of anxiety showed by their more anxious companions. To establish that this convergence reflected a contagious process, the researchers repeated the same experiment, but this time each individual vole

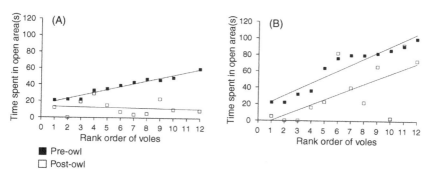

FIGURE 7.2 Contagion of fear in voles. Voles exposed as a group to attacks by an owl converged on the level of anxiousness shown by the more anxious voles, suggesting fear contagion in the group (A). Anxiousness was measured individually as the time spent in the open area of a maze. Individual voles were then ranked from the more anxious (less time in the open maze) to the less anxious voles. In a second experiment, each vole was subjected to attacks alone. As in (A), voles spent less time than before the attacks in the open area of a maze, but retained the same relative level of anxiousness, thus failing to show contagion of fear when isolated from one another. (*Adapted from Eilam et al. (2011)*).

was subjected to attacks alone. When retested, all voles tended to be more anxious as in the previous experiment, but the naturally less anxious voles remained noticeably less anxious than their more anxious companions. These results suggest that witnessing alarm responses by naturally more anxious companions was sufficient to induce more fear in less anxious voles.

In Sirot's model, the evolutionary stability of the rule used to weigh private and public information was not investigated, and as such it is not clear whether copying the vigilant state of neighbours represents a stable strategy. For instance, what would happen if one individual in the group adopted a slightly different rule from the others? Would it enjoy more or less success as a result? In general, testing how rare mutants perform in a population that has converged on a particular solution to a problem is needed to assess stability.

The ultimate reason for copying was left unspecified in the model. In other words, it is not clear how an individual would fare if it did not copy the vigilance state of neighbours. Intuitively, getting information from neighbours about predation risk is probably useful as any estimate gathered from many sources is typically closer to the true value (King and Cowlishaw, 2007). However, sampling the scanning behaviour of neighbours must take time and probably carries opportunity costs. In addition, neighbours may not always get it right, and copying their vigilance might lead to costly false alarms. Unless the fitness costs and benefits of copying are explicitly stated, it is difficult to determine the conditions under which this behaviour can evolve.

The contagion-of-fear model developed by Sirot (2006) is phenomenological; it prescribes copying without explaining why it occurs in the first place. Other phenomenological models have also considered the possibility that vigilance may not be independent amongst foragers in a group (Bahr and Bekoff, 1999; Beauchamp et al., 2012; Jackson and Ruxton, 2006). As with Sirot's model, their main purpose is to predict what happens if non-independent vigilance prevails in a group.

Empirically, a fruitful avenue for future research would be to establish whether hurried departures in a group tend to happen more frequently when the proportion of vigilant individuals is high, a situation where we would expect individuals in the group to be more nervous, and thus more likely to escape in a hurry.

7.2.2 Predator Targeting Behaviour

Synchronization of vigilance increases the amount of time during which no one in the group is vigilant. At those times, the ability to detect predators at the group level is temporarily compromised. Any reduction in overall vigilance appears at first sight counterproductive because seeking extra safety was thought to be one reason why individuals joined groups in the first place.

To understand how synchronization of vigilance might be adaptive, recall that direct detection of threats in a group, as opposed to indirect detection through collective detection, ensures a more rapid escape. When group members flee in alarm as the predator approaches, those who are left behind are more likely to be drawn from the pool of individuals that relied on collective detection rather than direct detection. A recent model argued that if predators preferentially target laggards, individuals in a group should be more likely to be vigilant when their neighbours are vigilant, and vice versa (Sirot and Touzalin, 2009). This extra vigilance when neighbours are vigilant decreases the relative risk of attack by reducing the odds of being left behind as the predator closes in. Conversely, if many neighbours are not vigilant, a decrease in vigilance makes sense since the risk of being targeted is now diluted with many equally vulnerable companions. Synchronization of vigilance occurs in this model because group members tend to copy the vigilance level of their neighbours. The model also showed that copying tends to produce temporal waves of vigilance with periodic rises and falls in the number of vigilant individuals in the group.

7.2.3   Empirical Evidence

Pulliam stated that finches, the species that inspired the first anti-predator vigilance model, appeared to raise their heads independently from one another, but provided no evidence (Pulliam, 1973). Independent vigilance amongst group members was statistically documented later in ostriches (Bertram, 1980), American pronghorns (Lipetz and Bekoff, 1982), house sparrows (Elcavage and Caraco, 1983), humans (Wawra, 1988), wild boars (Quenette and Gérard, 1992), doves (Cézilly and Keddar, 2012), degus (Quirici et al., 2008), and wombats (Favreau et al., 2009) (Box 7.1).

BOX 7.1 How to Detect Synchronization of Vigilance

In the earliest attempts to detect non-independent vigilance amongst group members, sequences of scans and inter-scan intervals were transformed into sequences of ones and zeroes, respectively, and correlated between pairs of individuals (Lipetz and Bekoff, 1982). Synchronization of vigilance, one form of non-independence, would produce a positive correlation between paired sequences. The hypothesis can be tested with a point-serial correlation, which is applicable to this kind of binary data. Alternatively, the likelihood of vigilance for one individual can be statistically related to the number of vigilant neighbours in the previous time interval (Pays et al., 2009b). Another statistical procedure uses time-series to uncover periodic rises and falls in the number of vigilant individuals in a group over time (Beauchamp, 2011).

Another approach can be applied to cases in which vigilance is recorded for all group members. In this case, it is possible to compare the observed proportion of times during which at least one group member is vigilant with the predicted proportion under the assumption of independent vigilance by all group members. To illustrate the calculation, consider a group of two foragers vigilant for proportions of time $p1$ and $p2$, respectively. If each individual scanned independently of the other, the predicted proportion of times where at least one group member is vigilant is given by: $1 - (1-p1) \times (1-p2)$. This can be compared to the observed proportion. Synchronization of vigilance would be suspected if the observed value is smaller than the predicted value.

Empirical research on synchronization faces significant methodological hurdles. Consider groups in which vigilance is high. In such groups, many individuals can be vigilant at the same time because each individual has a high probability of being vigilant. Because the expected value from independent vigilance is already high, it will prove very difficult to detect synchronized vigilance. By contrast, if vigilance tends to be low in the group there will be few opportunities to document synchronized scans. Generally, the largest difference between observed and predicted values will occur when vigilance is moderate in the group. As these examples show, lack of synchronization may be a consequence of low power to reject the null hypothesis of independent scanning. An index of synchronization independent of vigilance levels is needed.

Several recent studies provided evidence for synchronization of vigilance in birds and mammals (Beauchamp, 2009; Ebensperger et al., 2006; Fernández-Juricic et al., 2004c; Ge et al., 2011; Michelena and Deneubourg, 2011; Öst and Tierala, 2011; Pays et al., 2009a; Pays et al., 2009b; Pays et al., 2007a; Pays et al., 2007b; Pays et al., 2012b). In Eastern grey kangaroos, for instance, the observed proportion of times during which at least one group member was vigilant was lower than predicted under the null hypothesis of independent vigilance (Pays et al., 2007a), an indication that vigilance was synchronized (Fig. 7.3). In a further example, the number of sleeping gulls in a resting flock rose and fell periodically, as would be predicted if periods of high or low vigilance were synchronized (Beauchamp, 2011). None of the previous studies determined which of the two known processes underlying synchronization, namely, fear contagion or targeting of laggards, was more relevant.

One study failed to find evidence for synchronization of vigilance. In flocks of dark-eyed juncos, the addition of hungry birds, which were noticeably less vigilant, did not lead to a reduction in vigilance in the others (Lima, 1995a). Neighbours might have realized that low vigilance reflected hunger rather than low anxiety, which would explain the lack of copying. Nevertheless, the empirical evidence as a whole suggests that the independence assumption in traditional models of anti-predator vigilance is not always justified.

Obviously, vigilance may be synchronized in animal groups for more trivial reasons. Many individuals may be vigilant at the same time because each one independently detected the same external disturbance (Ruxton and Roberts, 1999). Ruling out external disturbances makes cases of synchronization more convincing. Synchronization of vigilance might also happen if foraging bouts tend to be synchronized. Because vigilance and foraging occupy a large part of the time budget of animals, any tendency by group members to start feeding at the same

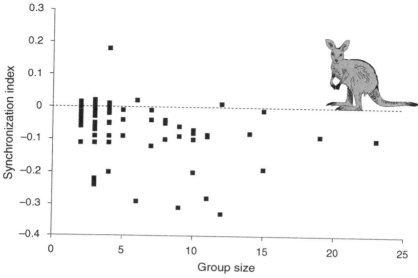

FIGURE 7.3   Synchronization of vigilance in Eastern grey kangaroos. The observed level of collective vigilance, defined as the proportion of times during which at least one individual was vigilant in the group, was systematically smaller than the predicted level based on independent vigilance by group members. The synchronization index represents the difference between observed and predicted values. Negative values indicate a tendency to synchronize vigilance. *(Adapted from Pays et al. (2007a)).*

time would synchronize vigilance as a side effect. Copying of foraging has been suspected in some studies (Fernández-Juricic and Kacelnik, 2004; Michelena and Deneubourg, 2011; Pays et al., 2012b).

Synchronization of vigilance requires monitoring the vigilance of other group members. Therefore, factors that make monitoring more difficult or more costly are expected to reduce the level of synchronization in a group. Synchronization might decrease in larger groups if monitoring many neighbours becomes too arduous (Lipetz and Bekoff, 1982). Synchronization might also be more difficult to achieve if individuals are further apart and thus more difficult to monitor. Periods during which no one is vigilant happen more frequently when vigilance is synchronized. This strategy may be quite dangerous in habitats with frequent disturbances. Consequently, synchronization might be less likely in more disturbed habitats.

Few studies have investigated these predictions. In red-necked pademelons, synchronization decreased as expected when individuals were further apart (Pays et al., 2009a). There was no effect of group size on synchronization in Eastern grey kangaroos (Pays et al., 2007a), Defassa waterbucks (Pays et al., 2007b), and American pronghorns (Lipetz and Bekoff, 1982). Synchronization decreased with group size in evening grosbeaks (Bekoff, 1995), but unexpectedly increased in common eiders (Öst and Tierala, 2011). In degu, a large South American rodent, synchronization was apparent at all group sizes but strongest in smaller groups (Ebensperger et al., 2006). In view of the divergent results obtained thus far, further tests of these predictions are needed.

7.3 COORDINATION OF VIGILANCE

Ideally, the ability to detect predators at the group level should remain constant. This could be achieved by coordinating vigilance so as to ensure that at least one individual is vigilant at all times. Coordination of vigilance represents a further example of non-independent vigilance. The use of sentinels is the simplest solution to coordinate vigilance. Sentinel behaviour will be covered in the following section. Here, I focus on groups without sentinels.

Coordination of vigilance appears quite rare in animal groups without sentinels. This was thought to reflect the prohibitive cost of monitoring the vigilance of neighbours (Ward, 1985). This might be true if monitoring neighbours took time away from foraging. As we saw in Chapter 3, prey animals can often scan and forage at the same time, suggesting that at least for such species, monitoring neighbours need not be too onerous. The degree of overlap between vigilance and foraging has emerged as a key factor in the evolution of vigilance coordination.

7.3.1 Why Coordinate Vigilance

A recent model explored the conditions that favour the evolution of coordination (Fernández-Juricic et al., 2004b). The model focused on two foragers subjected to random attacks and considered three outcomes following an attack (Fig. 7.4). First, the focal forager escapes after directly detecting the predator. Second, the focal forager is head down and fails to directly detect the approaching predator. However, the head-down forager can still escape if the neighbour detects the predator and rapidly passes along the information. This indirect way of escaping represents collective detection. Third, both foragers fail to detect the predator, in which case death looms. However, all is not lost for one lucky forager because the predator can only capture one each time. This escape route reflects predation risk dilution. Using these three different possible outcomes, the probability of capture can be calculated for the focal forager assuming that the two foragers scan with the same overall probability but independently from one another. For comparison, the probability of capture can also be calculated when the two foragers coordinate their vigilance. When coordinating their vigilance, only one of the two foragers scans at any one time (Fig. 7.4).

The model showed that coordination of vigilance, rather than independent vigilance, reduces the probability of capture if the probability of escaping through collective detection is greater than 0.5, the probability of capture predicted through risk dilution. In other words, coordination of vigilance evolves when the benefits from collective detection exceed the blind chances of surviving

Independent vigilance

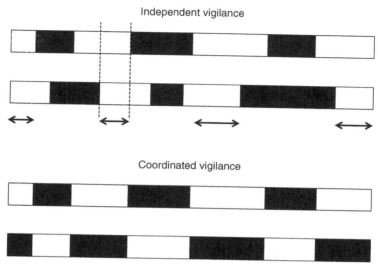

Coordinated vigilance

FIGURE 7.4   Coordination of vigilance. The advantage of coordinating vigilance is illustrated for two foragers both allocating half their time to vigilance. A scan is shown in black and inter-scan intervals are shown in white. When the two foragers scan independently of one another (upper panel), times during which both foragers scan simultaneously and times during which no one is scanning can occur frequently. Foragers are especially vulnerable to attacks when no one is scanning (shown with the arrows) because the probability of detection is small when not scanning. If both foragers fail to detect the approaching predator, one of the two foragers will be captured. When coordinating vigilance (lower panel), one of the two foragers is always alert and can escape easily when attacked. The non-vigilant companion can also escape, albeit with a lower probability dictated by the efficiency of collective detection.

an attack. When this happens, an individual can rely on collective detection to escape and should coordinate vigilance with the partner. The model also showed that coordination is even more beneficial when the probability of scanning is high and when the probability of detecting the predator when foraging is low.

The model did not consider genetic relatedness, but intuitively coordination should be more likely when individuals in the group are related. There would be more incentive to become vigilant when a neighbour is not if this neighbour is a close kin. Collective detection in this case could save the life of a kin rather than a potential competitor. An obvious next step would be to model the consequences of kinship for coordination. The idea that coordination might be more prevalent in kin groups was tested in one laboratory study recently, but there was no evidence for coordination of vigilance in related or unrelated pairs of degus (Quirici et al., 2008).

Another avenue for future research would be to determine whether a rare uncoordinated mutant in a coordinated group would achieve higher success than the others (Rodriguez-Girones and Vasquez, 2002). By curbing its vigilance, the mutant would benefit from the coordination of vigilance achieved by the others but obtain more resources. This exercise is important to assess the evolutionary stability of vigilance coordination.

7.3.2 Empirical Evidence

Few studies have investigated coordination of vigilance in animal groups without sentinels. Two recent studies conducted with pairs of animals feeding together failed to document coordination (Cézilly and Keddar, 2012; Quirici et al., 2008). This was particularly surprising in *Zenaida* doves, a species that apparently fulfils many of the conditions for coordination. Doves partially close their eyelids when feeding, making them less likely to detect approaching predators. Collective detection, however, is expected to be very effective because it relies on conspicuous acoustic cues of alarm (Cézilly and Keddar, 2012). Instead of relying on collective detection, by coordinating vigilance, doves in a pair were in fact more likely to initiate a scan at the same time, perhaps in response to external disturbances.

In three species of coral reef fish in Australia, by contrast, one member of a pair initiated vigilance more often when the other started to feed (Fox and Donelson, 2014), suggesting coordination of vigilance. In the three fish species, foraging amongst the reef matrix curtailed vision of the surroundings, which probably reduced the ability to detect predators when foraging. As we saw earlier, poor predator detection ability when foraging is a key factor in the evolution of coordination.

Mating pairs of white-tailed ptarmigan also coordinated their vigilance early in the breeding season (Artiss et al., 1999). In mating pairs of this species, females started to feed more often when their partner became vigilant. Similarly, feeding males became vigilant more often when their partner started to feed. Paired ptarmigans are not related, but each pair member has a vested interest in helping the other survive to successfully raise the family, which might foster coordination of vigilance. A detailed analysis of vigilance in mating pairs of other species might reveal unsuspected cases of coordination.

Common cranes in family groups also coordinated their vigilance (Ge et al., 2011). In the non-breeding season, common cranes typically forage in small family groups consisting of an adult pair with one or two offspring. Collective vigilance in each family was calculated as the proportion of time periods during which at least one parent was vigilant. The observed level of collective vigilance turned out to be larger than predicted under the assumption of independent vigilance by the two parents, suggesting that parent cranes tended to avoid periods during which no adult was vigilant (Fig. 7.5).

Coordination of vigilance makes sense for common cranes in this study. Cranes forage in small kin groups in a relatively open habitat. The close distance between family members probably facilitates collective detection. The level of vigilance is also quite high due to frequent disturbances by people. All these factors, as we saw earlier, can facilitate the occurrence of coordination.

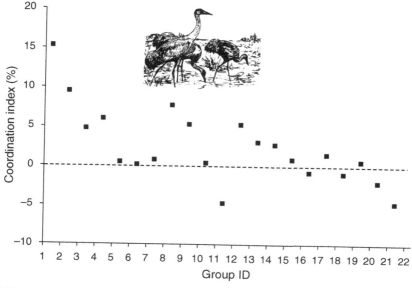

FIGURE 7.5 **Coordination of vigilance in common cranes.** In families of common cranes wintering in coastal China, the two parents coordinated their vigilance. The coordination index represents the deviation, expressed as a percentage, between the observed level of collective vigilance and the predicted level under the null hypothesis of independent vigilance. Positive values of the index indicate coordination of vigilance. *(Adapted from Ge et al. (2011)).*

Another situation where coordination of vigilance might take place is in a large group. Centrally located individuals can easily obtain indirect information about predation threats from their many neighbours on all sides ensuring effective collective detection. However, the probability of directly detecting a predator could be difficult because the very same neighbours disrupt sight lines to the outside. Peripheral individuals, on the other hand, benefit little from collective detection, but can detect predators more easily. Foragers in the middle of the group would benefit, I suspect, by coordinating their vigilance with the more vigilant peripheral companions.

7.4 SENTINEL BEHAVIOUR

Some of the earliest empirical investigations on animal vigilance described what appears to be sentinel behaviour. In California quail coveys, one individual typically maintained vigilance from a vantage point, which allowed the others to feed less warily nearby (Williams, 1903). In resting Chacma baboon troops, one or two individuals often sat on top of a rock to warn the others about possible danger (Elliott, 1913). Similar sentinel behaviour was noted in other early accounts of vigilance (Andrews and Naik, 1970; Balda and Bateman, 1971; Cary, 1901; Hall, 1960; Hardy, 1961; Westcott, 1969). Sentinel behaviour is now known in many species of birds and mammals, including jays and crows, babblers, parrots, weaver birds, social mongooses like meerkats, primates, and even one antelope species (Bednekoff, 1997).

In a true sentinel system, the number of sentinels fluctuates little over time despite the turnover of sentinels in the group. Low fluctuation reduces costly overlap in vigilance and maintains vigilance at a near constant level, which minimizes the amount of time during which no one is vigilant.

In a typical sentinel species, a small number of individuals maintain vigilance from a vantage point while the rest of the group performs other activities nearby like resting or foraging. Sentinels allow other group member to reduce their own vigilance, but rarely to the extent that it becomes absent. Spatial segregation of tasks and the fact that sentinel behaviour is mostly known from kin groups make it a special case of coordination. How coordination is achieved and how it benefits group members have been two topics of considerable interest to students of sentinel behaviour.

7.4.1   How is Coordination Achieved?

When all individuals forage in close contact, visual monitoring of neighbours is sufficient to ensure coordination. In a sentinel system, however, vigilant and non-vigilant group members may be quite far apart. How do group members assess the level of vigilance maintained by sentinels? Visual monitoring of the sentinels is probably onerous because it can take time away from foraging. Coordination of vigilance based on acoustic signals, however, would allow foragers to assess the level of vigilance of sentinels without interrupting foraging.

The watchman's song hypothesis suggests that sentinels use vocal cues to signal watchfulness to other group members without attracting the attention of nearby predators (Wickler, 1985). As evidence for the hypothesis, Wickler provided descriptions of sentinel calls seemingly, albeit not demonstrably used by various species to signal watchfulness. The most convincing evidence that sentinel calls play a role in vigilance comes from playback experiments. In a playback experiment, sentinel calls are broadcasted to the rest of the group under standardized conditions, which allows the researchers to pinpoint their effects on vigilance. In meerkats, probably the emblem of sentinel behaviour in the animal world, broadcasted sentinel calls, as opposed to white noise, reduced vigilance in the rest of the group. In addition, sentinel calls increased the degree of coordination in the group by reducing overlap in vigilance and decreasing the amount of time without sentinels (Manser, 1999) (Fig. 7.6). A similar sentinel calling system, which provides crucial information to nearby foragers, was documented in pied babblers (Bell et al., 2009). In babblers, vegetation obscures the view while foraging, making it difficult to detect predators without the aid of sentinels.

Calls provided by sentinels can thus signal their watchfulness. An alternative view is that calls are produced rather late in a sentinel bout to signal an

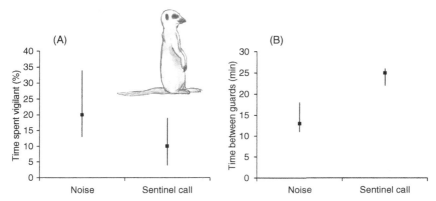

FIGURE 7.6 Sentinel behaviour in meerkats. Foragers in groups of meerkats spent less time alert after hearing broadcasted sentinel calls than background noises (A), suggesting that sentinel calls signal watchfulness. After sentinel calls, the time interval between guarding duties also increased (B). Black dots show the median response and bars extend from the 25th to the 75th percentiles. *(Adapted from Manser (1999)).*

impending shift in duties (Gaston, 1977). To distinguish between the two hypotheses, it is necessary to document the rate of calling during a sentinel bout. In Florida scrub-jays, calls provided by sentinels occurred quite irregularly during a sentinel bout and were not heavily concentrated near the end, rejecting both the aforementioned hypotheses (Bednekoff et al., 2008). While such calls undoubtedly provide valuable information to non-sentinel birds, additional cues must be used to coordinate vigilance in this species. Search for such cues is ongoing in this and other species as well.

Whatever the means used to coordinate vigilance, group members appear to compensate for the sentinel behaviour of others, indicating again that group members are aware of the watchfulness of sentinels. Recent studies showed that when some individuals spend more time as a sentinel, others compensate by spending less time as sentinels (Bednekoff and Woolfenden, 2006; Clutton-Brock et al., 1999; Wright et al., 2001b).

Another issue is how to prevent cheating in a sentinel system. An individual that foraged continuously without performing sentinel behaviour would benefit from the sentinel behaviour of others and get extra benefit from foraging longer. Vested interest in the survival of other group members probably prevents cheating. This may explain why sentinel behaviour is more likely in kin groups.

7.4.2   What Favours the Evolution of Sentinel Behaviour?

A naïve view of sentinel behaviour is that it has evolved to benefit the group. However, recent attempts at modelling sentinel behaviour view sentinel behaviour as the outcome of choices made by individual group members to maximize their own success.

While it is quite obvious how other group members benefit from the presence of sentinels, it is less clear how sentinels themselves benefit from their actions. Sentinels tend to occupy vantage points, which could, in fact, expose them to danger. Vantage points appear to be selected to provide a good view of the surroundings, which is necessary to detect approaching predators more easily. Vantage points in some species also tend to be close to protective cover (Clutton-Brock et al., 1999), although this is not always the case (Ridley et al., 2013). In some species, vantage points are adjusted according to the perceived risk of predation (Kern and Radford, 2014). Empirical evidence suggests that sentinels are more likely to detect threats than other group members (Manser, 1999; McGowan and Woolfenden, 1989; Rasa, 1987). However, evidence regarding the safety of sentinels is mixed: sentinels tend to be killed less often than foraging companions in some species (Clutton-Brock et al., 1999) but not in others (Ridley et al., 2013). Sentinels can see the predators first and initiate escape early, but lack of protective cover may be problematic in some species.

I described earlier the conditions that favour the evolution of coordination of vigilance. In particular, a low probability of detecting predators while foraging and effective collection detection both promote coordination. The choice of vantage points by sentinels fulfils both conditions. Sentinels can see well from these vantage points, but can also be seen just as easily by the rest of the group. In addition, alarm calls provided by sentinels can spread easily to the group. Coordination of vigilance makes sense under these conditions, but this discussion leaves open the details of how coordination is practically achieved. Some rules must exist to avoid having too many sentinels at the same time or none at all.

How do individuals achieve coordination without a leader to allocate sentry duty? A model suggested that it is clearly to the benefit of each individual to become a sentinel in an orderly fashion (Bednekoff, 1997). When the probability of detecting an approaching predator is low, foraging all at the same time increases risk. Becoming a sentinel provides safety but at the cost of losing foraging time. However, loss of foraging time may not be very costly if the forager is not very hungry anymore. Greater safety as a sentinel clearly can compensate for the decrease in foraging time as long as reserves are sufficiently high. If collective detection is effective and sentinels can easily detect predators, no more than one sentinel is needed at the same time. The model predicted that individuals with more reserves willingly choose to become sentinels and will rejoin the foraging group when hungry again. In addition, sentinel behaviour should be more likely when predation risk is higher.

These predictions have been tested extensively in Arabian babblers, a passerine bird species living in small territories in arid scrub of the Middle East. Sentinel behaviour in this species fitted the state-dependent predictions from the aforementioned model. Change-overs between sentinels occurred mostly when the current sentinel bird terminated its own bout (Wright et al., 2001a). There was little squabbling between birds exchanging duties, which suggests that there was no competition to occupy vantage points. The key experiment

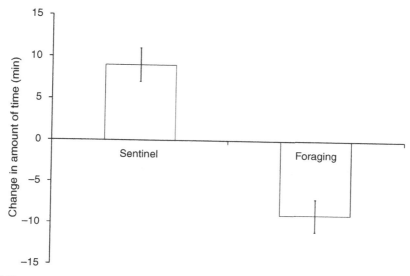

FIGURE 7.7 Sentinel behaviour in Florida scrub-jays. Individual birds that received supplemental food spent less time foraging and more time as sentinels than before supplementation, indicating that hunger affects sentinel behaviour. Mean and standard error bars are shown. *(Adapted from Bednekoff and Woolfenden (2003)).*

performed by the researchers addressed the relationship between the current state of a forager and sentinel behaviour. When targeted individuals in a group received supplemental food, those individuals were more likely to initiate sentinel behaviour (Wright et al., 2001b). An increase in reserves also increased sentinel behaviour in Florida scrub-jays (Bednekoff and Woolfenden, 2003) (Fig. 7.7). In other species, sentinel behaviour occurred more frequently when perceived predation risk increased (Kern and Radford, 2014; Ridley et al., 2010; Santema and Clutton-Brock, 2013). Altogether, these results strongly support the state-dependent model of sentinel behaviour. In cases in which sentinels are more exposed to predation, additional features must be invoked to explain coordination of vigilance like kin selection and cooperation (Ridley et al., 2013).

7.5 CONCLUSIONS

There are clearly good reasons to pay attention to the vigilance behaviour of neighbours. Their vigilance can provide valuable information about current predation risk, which can supplement personal information. Copying the vigilance of neighbours could be a life-saving tactic if it reduces the chances of being left behind when a predator targeting laggards comes along.

Individuals can also monitor the vigilance of others to avoid being vigilant at the same time, thus ensuring a more constant surveillance at the group level. Coordination of vigilance can lead to a spatial segregation of tasks within the group: some individuals maintain vigilance from vantage points while others forage nearby. Even though early vigilance models all assumed independent scanning amongst group members, synchronization and coordination of vigilance suggest that independent scanning might actually be the exception that requires a special explanation.

Synchronization of vigilance has been documented in several species, but several questions remain unanswered. Two different mechanisms, namely, fear contagion and predator preference for laggards predict the occurrence of synchronization. A promising avenue for future research would be to investigate vigilance in a situation in which only the premises of one model are valid. In a prey animal species where laggards are not targeted by predators, for instance, synchronization of vigilance is more likely to reflect fear contagion.

We need more direct evidence that prey animals use the vigilance of neighbours to get a better estimate of predation risk, the necessary condition for the contagion-of-fear model of synchronization. Simple choice experiments come to mind to address this issue experimentally. When given a choice between two groups of equal sizes, prey animals should prefer to join the less vigilant group because low vigilance betrays safer conditions. Adding more fearful individuals in a group should increase vigilance in the others. Fearfulness could be induced experimentally by the chemical means identified in Chapter 3. Alternatively, it could be possible to introduce naturally more fearful individuals into groups as was the case in voles.

Coordination of vigilance is rare in non-sentinel species. I have identified two contexts in which such coordination might be fruitfully investigated in the future, namely, in mating pairs and in large groups. Perhaps a more fundamental reason why coordination is less likely in groups without sentinels is that these groups only form on a short-term basis. In species with fluid groups, a well-fed individual might simply leave the group for safety elsewhere rather than become a sentinel for the remaining group members. Repeated interactions with the same companions might be needed to establish trust and increase the value of coordination.

In sentinel species, it is fairly easy to see how coordination can arise if sentinels enjoy greater safety and pay only a small foraging cost. But there is still relatively little evidence that sentinels are safer than other group members and that sentinel behaviour is state-dependent. If coordination of vigilance requires more than mutualism between selfish sentinels, some form of cooperation appears necessary. In this case, it would be interesting to examine whether sentinels provide false information to the rest of the group to increase their own benefits. Recent work showed how sentinel calls can sometimes be used deceptively (Baigrie et al., 2014). Such considerations take us quite far from the simple expectation that sentinels honestly signal their watchfulness to other group members.

Chapter 8

Applied Vigilance

8.1 INTRODUCTION

What would a typical vigilance study look like? If it were possible to distil the information from the thousand-plus studies on vigilance and express the contents into a few words, this is what I think we would get. A typical vigilance study investigates how an ecological factor, most likely sex or group size, influences the allocation of time to vigilance in a diurnal species, such as a house sparrow or an elk, forming groups with conspecifics, foraging in an undisturbed habitat, and facing significant threats from predators. In this last chapter, I explore atypical vigilance studies in the hope of fostering research on vigilance outside this conventional box.

Unconventional studies of vigilance take place in neglected contexts. There are several reasons behind the neglect; some are practical and others historical. For practical reasons, most studies of vigilance take place during the day. Observations at night require specialized equipment and non-negligible sacrifices in sleep time. Nevertheless, many species forage at night (McNeil et al., 1992), a period characterized by low light levels. Because predator detection and mutual warning in a group often rely on vision, what happens to vigilance at night certainly has the potential to shed much light on the proximate control of vigilance. Again for practical reasons, most vigilance studies are conducted in areas close to research institutions. However, some far-flung locales offer unique opportunities to investigate vigilance in atypical circumstances. As an example, some isolated islands are quite devoid of predators (Quammen, 1996). This provides a golden opportunity to explore what happens to vigilance when predation risk is negligible. If vigilance is mostly an adaptation to avoid predation, relaxed predation pressure should have a strong impact on vigilance unless vigilance serves other purposes.

Vigilance has typically focused on wild species foraging in their natural habitats reflecting the legacy of ethology, the scientific discipline that focuses on animal behaviour (Tinbergen, 1951). Nevertheless, examining atypical species can be illuminating for several reasons. Species like us or domesticated animals often face little predation pressure, which allows us to examine how vigilance fares in a novel environment. Encroachment of natural habitats by human activities also provides an opportunity to examine how species react to human disturbances. Vigilance, in this context, can provide a very useful tool

Animal Vigilance. http://dx.doi.org/10.1016/B978-0-12-801983-2.00008-5

to determine the level of stress caused by disturbances, and ultimately to assess how anthropogenic changes are perceived by animals.

Vigilance has typically focused on single-species groups. However, many species form mixed-species groups (Diamond, 1981). The occurrence of different species in the same group can broaden predator detection ability by bringing together senses that are tuned differently. The addition of such species could decrease the need for vigilance against predators. However, other species can also create headaches for vigilance. For instance, some species specialize in stealing food from others, which may force targeted species to monitor neighbours as well as predators.

In the following, I focus first on how animals adjust vigilance to human disturbances and then explore vigilance in various atypical contexts including when predation risk is reduced and in mixed-species groups.

8.2 VIGILANCE AND FLIGHT INITIATION DISTANCE

Flight upon a disturbance is a common response by animals under threat. The distance at which an animal flees from an approaching threat has been of considerable interest to both animal behaviour students and conservation biologists. Early research on flight initiation distance (Altmann, 1958; Walther, 1969) aimed at uncovering factors that govern how quickly a species, typically a big game animal, takes flight upon disturbance. The idea was that flight initiation distance could be used as a proxy for vulnerability: species that flee earlier than others probably face a higher predation risk. Factors that influence flight initiation distance can thus shed light on the perception of predation risk. More practically, information about flight initiation distance is important for conservation biologists to set the size of buffer zones to minimize the impact of human activity on wildlife.

In a typical empirical study of flight initiation distance, a person walks towards a target animal at a set pace. Because approaches by real predators are typically rare and difficult to witness, standardized approaches with human subjects have been used instead as a proxy for predation threats. The distance at which the target animal first reacts is known as the detection distance while the distance at which flight occurs is known as the flight initiation distance (Stankowich and Blumstein, 2005; Ydenberg and Dill, 1986). Vigilance can influence both distances.

Detection for an ever-vigilant animal would occur as soon as the senses pinpoint the source of the threat. In this ideal scenario, detection would be expected to vary from species to species reflecting their ability to extract threat signals from the noisy habitat. Avian species with larger eyes, for instance, might be able to visually detect a threat earlier because of their greater visual acuity (Fernández-Juricic et al., 2004a). Species that generally spend more time scanning could also detect threats earlier (Fernández-Juricic and Schroeder, 2003). In a visually cluttered habitat, visual threats signals can be masked by vegetation and latencies to detection should be longer (Devereux et al., 2006). Similarly, an increase in ambient noise levels might reduce the ability to detect sounds generated by an approaching predator leading to slower detection (Quinn et al., 2006).

The ever-vigilant animal would probably soon die of starvation. Allocation of time to other activities must reduce the ability to detect threats and initiate early flight. Sleeping prey animals, for instance, can be approached undetected more easily (Lima et al., 2005). Foraging is also known to interfere with the detection of threats by blocking sightlines or producing noises (Kaby and Lind, 2003; Lima and Bednekoff, 1999a; Lynch et al., 2015). Even time spent vigilant against predators can be diverted by rival sources of attention like competitors. Summing up, detection distance will reflect the interplay between sensory processing ability, habitat noise, and the amount of time animals can allocate to predator detection. Factors that influence investment in anti-predator vigilance will likely affect the ability to detect predators early and indirectly flight initiation distance.

The above discussion seems to imply that prey animals that invest less time in anti-predator vigilance should be less likely to detect threats early, which means that they will tend to flee at shorter distances. However, recent work suggests that this simple conclusion is not necessarily true. Initially, competing activities such as foraging and sleeping were viewed as being entirely incompatible with vigilance. But research showed that some vigilance can be maintained even when animals are not overtly vigilant. Sleeping animals peek periodically during sleep and are able to maintain a minimum level of vigilance (Rattenborg et al., 1999b). Animals with laterally placed eyes can monitor areas directly above them even when foraging head down (Fernández-Juricic et al., 2008; Lima and Bednekoff, 1999a; Wallace et al., 2013). Other species can monitor their surroundings while searching or handling food in a head-up posture (Fortin et al., 2004b; Kaby and Lind, 2003). The consensus now is that some level of vigilance can be maintained by non-overtly vigilant animals, but their ability to detect threats is probably weaker (Bednekoff and Lima, 2005; Lima and Bednekoff, 1999a). Species that are able to maintain some vigilance when non-overtly vigilant might be expected to show longer flight initiation distances. Fundamental information about the level of vigilance animals can maintain during different activities is therefore essential to predict flight initiation distance.

Any factor that reduces time spent overtly vigilant has the potential to influence flight initiation distance. Many drivers of vigilance are known (Chapter 4), but chief amongst them is group size. While an increase in group size typically leads to a decrease in individual allocation of time to vigilance, corporate vigilance in a group tends to increase in larger groups. As a consequence, predator detection should occur sooner in larger groups with a concomitant increase in flight initiation distance. In a meta-analysis involving a wide range of animal taxa, flight initiation distance tended, as predicted, to be longer for prey animals in larger groups (Stankowich and Blumstein, 2005).

For prey animals in groups, the level of corporate vigilance is key to understanding the ability to detect predators early. Recent developments in animal vigilance research have shown that corporate vigilance, however, is not necessarily a fixed property within the same group. In a group, fluctuation in corporate vigilance over time is expected in two contexts. First, we saw earlier that corporate vigilance is predicted to show cyclical ups and downs when individuals copy the vigilance of one another (Beauchamp et al., 2012; Sirot and Touzalin, 2009). Second, a decrease in corporate vigilance with time is predicted when prey animals face observant predators (Beauchamp and Ruxton, 2012a) or if they assess predation risk in a dynamic fashion (Sirot and Pays, 2011). Such temporal fluctuations imply that the ability to detect predators, and indirectly flight initiation distance, could vary from time to time within the same group.

In addition to group size, many factors are known to influence the amount of time allocated to vigilance. Here, I discuss how food density could influence flight initiation distance through its effect on vigilance. Samples of flight initiation distances may be taken from areas that differ markedly in resource availability and quality. If animals face time constraints, the level of vigilance is expected to decrease as the rate of food intake increases (Chapter 4). In rich food patches, flight initiation distance might be shorter because of the expected reduction in vigilance. Lack of control over food density, or any other factor known to influence vigilance, can affect the accuracy of flight initiation distances measured in the field. Generally, vigilance can have profound effects on flight initiation distance in animals.

8.3 VIGILANCE IN DISTURBED HABITATS

Studies on flight initiation distance investigate the response of one animal to one standardized disturbance caused by one person. Encroachment of natural habitats by humans causes a much wider range of disturbances to wildlife. Disturbances caused by human activities can lead to changes in habitat use and behaviour in animals (Boyle and Samson, 1985; Chace and Walsh, 2006; Gill et al., 1996). While most human activities pose no immediate threat to survival, animals often perceive human-caused disturbances as a form of predation risk, responding to stimuli associated with human presence as if they were a real source of danger akin to predators (Frid and Dill, 2002). Loud noises caused by airplanes or the intrusion of well-intentioned animal behaviour students in a territory send prey animals scurrying to safety. Further to the view that human activities are perceived by animals as a form of predation risk, animals are predicted to respond more strongly when disturbances are more frequent and closer (Beale and Monaghan, 2004b).

How do we evaluate the impact of human disturbances on animals? Direct indicators like mortality or reproductive success are probably best, but require arduous long-term monitoring (Gill, 2007; Tarlow and Blumstein, 2007). Indirect indicators can be more easily obtained, but often lack sensitivity in detecting an association between disturbances and behaviour. If disturbances caused by humans are indeed viewed as a form of predation risk, vigilance might be a good indirect indicator of the amount of stress caused by disturbances. Increased vigilance might also be a link with more direct impacts of disturbances since it can influence the amount of time allocated to feeding and ultimately fitness.

Do human disturbances alter vigilance in animals? I compiled cases from the literature for both captive and wild animals. I only included cases that contrasted vigilance in a disturbed and undisturbed situation (Table 8.1). I also distinguished two types of studies. In the first type, researchers documented responses by animals to a fixed number of disturbances. One example would be responses to a single canoe passing by a breeding colony. Here, the distance to the source of disturbance can vary, but not the number of disturbances. In other cases, researchers compared vigilance responses in disturbed and undisturbed habitats assuming a fixed level of disturbances in the disturbed habitat. In the second type, researchers noted responses to the number of disturbances. One example would be responses to the varying number of cars passing by a road during a fixed time interval.

TABLE 8.1 Effect of Human-Caused Disturbances on Vigilance in Captive and Wild Animals

Species	Effect on vigilance*	Source of disturbance	References
Captive animals: Fixed number of disturbances			
12 species of ungulates	I	Zookeeper; effect more marked for females and in large species	Thompson (1989)
Captive animals: Variable number of disturbances			
Mandrill	I	Zoo visitors	Chamove et al. (1988)
Orangutan	I	Zoo visitors	Birke (2002)
Père David's deer	I	Pen visitors	Li et al. (2007)
Koala	I	Zoo visitors	Larsen et al. (2014)
Wild animals: Fixed number of disturbances			
Elk	N	Car traffic	Schultz and Bailey (1978)
Brent goose	N	Farmers	White-Robinson (1982)
Pronghorn	I	Human presence	Berger et al. (1983)
Piping plover	I	Pedestrians	Flemming et al. (1988)

(Continued)

TABLE 8.1 Effect of Human-Caused Disturbances on Vigilance in Captive and Wild Animals *(cont.)*

Species	Effect on vigilance*	Source of disturbance	References
Brent goose	N	Pedestrians, vehicles	Riddington et al. (1996)
Trumpeter swan	I	Aircraft, truck, or pedestrian	Henson and Grant (1991)
American crow	I	Not clearly specified; effect more marked for closer disturbances	Ward and Low (1997)
Waterfowl	N	Military aircraft	Conomy et al. (1998)
Emperor penguin	I	Helicopter	Giese and Riddle (1999)
Sandhill crane	I	Car; reacted more strongly closer to the road	Burger and Gochfeld (2001)
Southern elephant seal	I	Researchers	Engelhard et al. (2002)
Dall's mountain sheep	I	Helicopter	Frid (2003)
Dall's mountain sheep	I	Nearby researchers; strongest reaction in adults	Loehr et al. (2005)
Brown bear	I, N	Tourists; stronger reaction in males than in females with cubs	Nevin and Gilbert (2005)
Coot	I	Dog barks	Randler (2006a)
California ground squirrel	I	Wind turbine	Rabin et al. (2006)
Brown bear	I	Tourists; more reaction closer to tourists	Rode et al. (2006)
Three species of penguins	I	Pedestrian	Holmes (2007)
Feral goat	I	Helicopter; reacted more strongly when helicopter was closer	Tracey and Fleming (2007)
Black-crown night heron	I	Canoe or pedestrian; responded more strongly to inquisitive movement	Fernández-Juricic et al. (2007b)
Olympic marmot	I	Cars, hikers	Griffin et al. (2007)
Andean condor	I	Distance from roads	Speziale et al. (2008)

TABLE 8.1 Effect of Human-Caused Disturbances on Vigilance in Captive and Wild Animals *(cont.)*

Species	Effect on vigilance*	Source of disturbance	References
House finch	D	Urbanization; less vigilance in more urbanized areas perhaps related to the presence of fewer predators	Valcarcel and Fernández-Juricic (2009)
Great bustard	I	Car traffic, pedestrians; responded more strongly to pedestrians	Sastre et al. (2009)
Spotted hyena	I, N	Responded to livestock and bell sounds but not to tourists	Pangle and Holekamp (2010a)
Black-tailed prairie dog	I	Urbanization	Ramirez and Keller (2010)
Red-crowned crane	I	Reserve visitors, farmers	Ge et al. (2011); Li et al. (2013); Wang et al. (2011)
Black-tailed prairie dog	N	Urbanization	Magle and Angeloni (2011)
Eurasian coot	N	Human presence	Severcan and Yamac (2011)
Magpie-lark	N	Urbanization	Kitchen et al. (2011)
Tibetan antelope	I	Road traffic	Lian et al. (2011)
Elk, Pronghorn	I	Pedestrian, bicycle	Brown et al. (2012)
Woodchuck	N	Urbanization	Lehrer et al. (2012)
Nubian ibex	I	Hikers; stronger response when hikers blocked escape route	Tadesse and Kotler (2012)
Piping plover	I	People and vehicles	Maslo et al. (2012)
Mediterranean mouflon	I	Hunters, recreational activities	Benoist et al. (2013)
Blue sheep	I	Pedestrians, tour buses; responded more strongly to pedestrians	Jiang et al. (2013)
Common mynah	N	Urbanization; fled more readily in urban centres but no adjustments in vigilance	McGiffin et al. (2013)

(Continued)

TABLE 8.1 Effect of Human-Caused Disturbances on Vigilance in Captive and Wild Animals *(cont.)*

Species	Effect on vigilance*	Source of disturbance	References
Père David's deer	I, N	Visitors; responses in males only	Zheng et al. (2013)
Shorebirds	I	Cars, buses	Schlacher et al. (2013)
Prairie dog	I	Road traffic	Shannon et al. (2014a)
Elk, Pronghorn	D	Road traffic; lower predation risk in disturbed habitats	Shannon et al. (2014b)
Kentish plover	I	Pedestrians	Martín et al. (2014)
Magellanic penguin	I	Tourists	Villanueva et al. (2014)
Wild animals: Variable number of disturbances			
Snow goose	N	Hunters, aircrafts	Bélanger and Bédard (1990)
Piping plover	I	People on the beach	Burger (1991, 1994)
Sanderling	I	People on the beach	Burger and Gochfeld (1991a)
Caribbean flamingo	I	Tour boats	Galicia and Baldassarre (1997)
Shorebirds	I	People on the shore; reacted more strongly to speedier movements	Fitzpatrick and Bouchez (1998)
Waterbirds	I	Visitors; reacted more strongly to noise than number of visitors	Burger and Gochfeld (1998)
Woodland caribou	I	Ecotourists	Duchesne et al. (2000)
European blackbird	I	Pedestrians in a park	Fernández-Juricic and Telleria (2000)
Desert bighorn sheep	N	Hikers, vehicles, bikers; more disturbances did not elicit more responses	Papouchis et al. (2001)
Mountain gazelle	I	Hikers and off-road vehicles; strongest reaction in larger groups	Manor and Saltz (2003)
New Zealand dabchick	I	Motor boats	Bright et al. (2003)

TABLE 8.1 Effect of Human-Caused Disturbances on Vigilance in Captive and Wild Animals *(cont.)*

Species	Effect on vigilance*	Source of disturbance	References
Polar bear	I, N	Tourist vehicles; more vehicles did not elicit a stronger response; females decreased their vigilance	Dyck and Baydack (2004)
American oystercatcher	I	Boats; strongest reaction when predation risk was high	Peters and Otis (2005)
Whopper swan	I	Pedestrians, farmers	Rees et al. (2005)
Pronghorn	I	Car traffic; strongest reaction close to the road	Gavin and Komers (2006)
Elk, bison	I	Snowmobiles	Borkowski et al. (2006)
Australian sea lion	N	Visitors on the beach	Orsini et al. (2006)
Red deer	N	Hikers, bicycles; increased vigilance regardless of the number of disturbances	Jayakody et al. (2008)
Elk	I	Car traffic	Clair and Forrest (2009)
Yellow-bellied marmot	I, N	Car traffic, bicycles, pedestrians; responded more strongly to cars and bicycles than pedestrians; responses stronger in adults than juveniles	Li et al. (2011b)
Elk, Pronghorn	N	Car traffic noises	Brown et al. (2012)
Elk	I, N	Car traffic; more responses closer to roads; no effect of pedestrians or bicycles	Ciuti et al. (2012)
Harbour seal	I	Seal watchers	Granquist and Sigurjonsdottir (2014)

*I = increase; N = no effect.

Adjustments in vigilance to human-caused disturbances have been investigated in many species, ranging from small passerine birds to large predators like the polar bear. Disturbances include a large array of stimuli, from hikers on a trail to tour boats and car traffic. Disturbances involve both visual (e.g. people walking on a trail) and auditory stimuli (e.g. traffic noise). Perusing the vast literature on the subject, it became abundantly clear that vigilance in animals tends to increase when disturbances are more frequent and closer. This was true for captive as well as for wild animals. There are, of course, exceptions to the above patterns, and these are explored below.

Disturbances often involve anthropogenic noises. Noise can distract animals and potentially mask sounds produced by approaching predators (Barber et al., 2010; Chan et al., 2010). In many cases, visual and auditory stimuli are entwined, which makes it difficult to isolate the disturbing features. As an example, large groups of people walking by are visually more conspicuous, but also likely noisier. Noise alone could explain an adjustment in vigilance (Quadros et al., 2014). A recent study explored this issue in caged koalas. Not surprisingly, koalas increased their vigilance when the number of visitors increased (Larsen et al., 2014) (Fig. 8.1). But koalas exposed to the sounds produced by visitors, but not their sights, showed a similar increase in vigilance, which suggests that part of the vigilance response was related to noises made by visitors. In several species of waterbirds in a Florida reserve, changes in vigilance correlated better with the amount of noise produced by visitors than to their numbers (Burger and Gochfeld, 1998). Noise alone is certainly sufficient to induce changes in vigilance.

Not all noises seem to produce the same type of responses in animals. In addition, different species often react differently to the same noises. Large ungulates foraging close to roads showed little reaction to car traffic noises

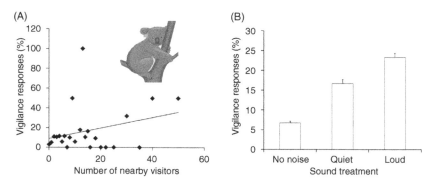

FIGURE 8.1 **Human disturbances and vigilance.** (A) Captive koalas showed more vigilance responses when more visitors passed by their enclosures. (B) Broadcasts of noisier visitors also triggered more vigilance responses. Means and standard error bars are shown. *(Adapted from Larsen et al. (2014)).*

(Brown et al., 2012; Schultz and Bailey, 1978). By contrast, broadcasted car traffic noises increased vigilance in prairie dogs (Shannon et al., 2014a). Many features of noise are likely to play a role, including frequency and volume. Sudden loud noises, for example, often elicited strong responses in animals (Pangle and Holekamp, 2010c; Tracey and Fleming, 2007) while constant loud noises produced little effect as in the above study with ungulates. Which combinations of noise parameters produce more responses, and why different species react differently remain largely unexplored.

Responses to urbanization have been recorded in many species. Urbanization encompasses a wide range of human-caused disturbances not all of which are quite understood. Because human-caused disturbances are presumably more frequent and/or closer in more urban areas, the simple expectation would be that vigilance should increase with urbanization. Surprisingly, many studies revealed quite the opposite. Reduced numbers of predators in urban settings, or in disturbed areas in general (Crooks, 2002; Dyck and Baydack, 2004; Leighton et al., 2010; Shannon et al., 2014b; Valcarcel and Fernández-Juricic, 2009), could actually decrease the need for vigilance. Another unexpected effect was documented in brown bears exposed to viewing tourists. In brown bears, feeding females monitor large, potentially aggressive males to avoid encounters. But because large males tended to avoid areas with tourists, females actually relaxed their vigilance there (Nevin and Gilbert, 2005). Counterintuitively, disturbances can create a spatial and temporal refuge from predators and foes, which can decrease the need for vigilance. In addition to differences in predation or social risk, changes in group size or food availability along an urbanization gradient should also be considered when interpreting adjustments in vigilance (Anderies et al., 2007).

Many studies suggested that animals can adapt to human-caused disturbances (Ikuta and Blumstein, 2003; Knight, 2009; Schultz and Bailey, 1978; Stankowich, 2008). In one striking example, black-tailed prairie dogs maintained less vigilance in urban settings than in their native grassland habitats. This was especially the case for older colonies exposed longer to disturbances, which the researchers interpreted as a form of habituation (Magle and Angeloni, 2011). Predictability of non-threatening disturbances has been viewed as a major factor contributing to habituation. Prairie dogs might have learned over time that some disturbances are innocuous. As another example, tour buses passing by regularly and at slow speed elicited fewer responses from blue sheep in a Chinese reserve than pedestrians that tended to arrive at unpredictable times and locations (Jiang et al., 2013). Habituation can reduce the impact of human disturbances on vigilance in animals.

Habituation represents a cohort phenomenon: exposure to relatively constant and benign stimulation allows individuals within a cohort to reduce their vigilance over time. Lack of effect of disturbances could alternatively be attributed to some form of selection whereby animals with the lowest responses to disturbances remain in disturbed areas while others simply emigrate or die. To test this hypothesis requires following a cohort of marked individuals over time to document differential migration or mortality.

Increased vigilance in response to disturbances can be viewed as both a reactive and proactive behaviour. As a proactive behaviour, vigilance increases the chances of detecting future threats early. As a reactive behaviour, vigilance can help assess the level of threat associated with the current disturbance. Threat assessment in this situation resembles predator inspection behaviour. During inspection, prey animals use vigilance to assess the level of motivation of a predator and the likelihood of attack (FitzGibbon, 1994; Leuthold, 1977). Reactive vigilance would be useful to avoid fleeing needlessly when there is little risk associated with a disturbance.

An alternative explanation for increased vigilance in reactive situations was proposed for captive animals exposed to visitors. Increased vigilance in this situation allegedly betrays a desire for novel social interactions (Morris, 1964). However, captive animals might have learned that humans are associated with food or companionship, which would be sufficient to explain approaches and curiosity. Wild animals typically flee when humans come too close, suggesting that neophilic responses are probably rare in the wild. Nowadays, most studies with captive animals consider the presence of visitors as a source of stress rather than enrichment (Claxton, 2011; Davey, 2007).

In addition to direct disturbances, human activities can also cause indirect changes in perceived risk. Many species are known to increase their vigilance during the hunting season or in areas where hunting is allowed (Benhaiem et al., 2008; Benoist et al., 2013; Casas et al., 2009; Ciuti et al., 2012; Crosmary et al., 2012; Jayakody et al., 2008; Pauli and Buskirk, 2007; Sonnichsen et al., 2013). Previous encounters with hunters and noises caused by guns are probably sufficient to increase the perception of risk and alter vigilance levels. Similarly, animals might also learn that some parts of the habitat are associated with a greater potential for disturbances by people. As a case in point, Pyrenean chamois increased their vigilance when closer to hiking trails (Lamerenx et al., 1992). Proactive changes in vigilance in response to the threat of disturbances might have a greater impact on animals than the rarer responses to actual threats.

8.4 VIGILANCE WHEN PREDATION RISK IS RELAXED

Predation avoidance is one of the main reasons why prey animal species invest time and energy in vigilance. Models all predict an increase in vigilance in riskier situations. If vigilance levels mostly track changes in predation risk, animals should exhibit little vigilance in the absence of predation risk. The occurrence of vigilance when predation risk is negligible would suggest that factors other than predation risk, such as competitors, sustain current vigilance or that vigilance has yet to be weeded out by natural selection. Here, I explore what happens to vigilance when predation risk is negligible.

8.4.1   Vigilance in Humans

Most human populations have not been exposed to any significant predation pressure for generations. What happens to vigilance in modern human populations is thus of considerable interest if vigilance is viewed as a response to predation risk.

In modern settings, vigilance in humans has been thought to reflect social monitoring rather than predator detection (Dunbar et al., 2002). Based on this premise, several predictions can be made about the occurrence of vigilance in different social contexts. Vigilance might be used to look for friends, a tendency that should be stronger when people are alone or in small groups. Vigilance might also be used to detect potential mates, in which case vigilance should vary primarily with group composition rather than group size. If people are seeking partners of the opposite sex, vigilance should be higher in single-sex groups than in mixed-sex groups. In addition, in mixed-sex groups, vigilance should decline with the number of opposite sex companions. Researchers tested these predictions in gathering places where casual groups of adult men and women form spontaneously. The findings revealed that both men and women decreased their vigilance in larger groups, but women in general tended to be less vigilant than men (Dunbar et al., 2002) (Fig. 8.2). Vigilance also varied with group composition. After sifting through the evidence, the researchers found the best support for the mate-searching hypothesis. The lower vigilance by women probably reflected an effort to avoid attracting unwanted attention.

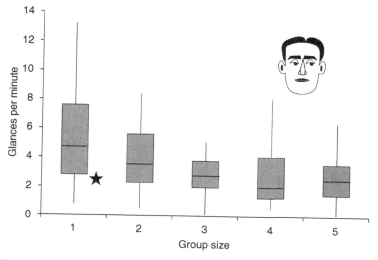

FIGURE 8.2   **Vigilance in humans.** Glance frequency by men in a cafeteria decreased with group size. The star shows the average glance rate of women alone. Box plots show the median glance frequency and the inter-quartile range. The box and extending bars cover 95% of the data. *(Adapted from Dunbar et al. (2002)).*

By contrast, an earlier study with human subjects related the decrease in vigilance with group size to the ghost-of-predation-past theory (Wirtz and Wawra, 1986). This theory posits that traits, such as vigilance, that are no longer useful can nevertheless persist in a population if the traits are not too costly or if the selection regime only changed recently (Lahti et al., 2009). Vigilance, according to this theory, evolved long ago when humans were vulnerable to predators, but persists nowadays despite the reduction in predation pressure. The once adaptive decrease in vigilance with group size thus continues to this day in response to predator ghosts. The Dunbar et al. study, however, reminds us that vigilance may be targeted at real companions rather than predator ghosts. That social function alone may be sufficient to maintain vigilance despite the reduction in predation risk.

In addition to the effect of group size, other aspects of vigilance are amenable to study in humans. In particular, the tendency to copy vigilance, which has been documented in many species of birds and mammals (Chapter 7), may be highly relevant in a species like ours in which decision-making is often influenced by the behaviour of others (Danchin et al., 2004). Vigilance copying was examined in a recent laboratory study in which human subjects were financially rewarded for solving mathematical problems (Kameda and Tamura, 2007). Focusing on the problems at hand, however, hampered the ability to detect signals in the room. Missing out on the signals reduced the payoffs. This game thus simulated the typical trade-off between vigilance and foraging payoffs faced by animals in groups. As expected, individuals allocated more time to vigilance as the costs of failing to detect the signal increased. But the more interesting finding is that when the perception of risk increased, vigilance reached a much higher level than predicted, which suggests that conformity in behaviour overruled individual interests. This tendency to copy the vigilance of neighbours mirrors the findings that animals in groups often synchronize their vigilance (Chapter 7). Recent studies also suggest that people are sensitive not only to the vigilance state of their neighbours but also to more subtle cues of vigilance including gaze direction (Gallup et al., 2012) and associated facial expressions (Gallup et al., 2014).

Humans have forward-facing eyes. This feature makes it relatively easy to establish the target of vigilance, which would be particularly useful to determine whether vigilance is aimed at companions in the group rather than potential outside threats. Human research on vigilance from an adaptive point of view certainly deserves more attention. The above results suggest that vigilance in humans serves at least a social monitoring function.

8.4.2 Vigilance in Domesticated Animals

Domesticated animals have also been used to examine vigilance in species that currently face little predation pressure. The obvious question is what becomes of vigilance in animals domesticated thousands of generations ago. Although domestication can lead to dramatic changes in appearance and behaviour, researchers have argued that domesticated species may have retained their basic social characteristics (Price, 1984), in which case vigilance would still be influenced by the many drivers associated with predation risk.

To examine this issue, vigilance in many domesticated species has been related to group size, one of the main drivers of vigilance. One study with free-ranging cattle showed that individuals reduced their vigilance in larger groups (Kluever et al., 2008), suggesting a retention of appropriate risk management despite domestication. Similarly, grazing dairy cows maintained in smaller groups moved their heads more frequently and spent more time ruminating head-up, suggesting that they perceived a higher risk in smaller groups (Rind and Phillips, 1999). Reindeer, a domesticated form of caribou, also decreased their vigilance in larger groups (Reimers et al., 2011). Vigilance in domestic sheep decreased with group size when controlling for food density (Michelena et al., 2012) (Fig. 8.3). Time spent grazing in sheep also increased with group size, suggesting group-size effects on vigilance (Penning et al., 1993). An effect of group size on vigilance has also been documented in chickens (Newberry et al., 2001).

Studies with domesticated animals often take place in confined spaces, such as enclosures or pens. This can be problematic because in an area of fixed size any increase in group size immediately changes the density of animals (Christman and Leone, 2007). As we saw in Chapter 4, animal density can exert a strong influence on vigilance, independently of group size. To avoid this issue, animal density can be maintained constant by providing more space to individuals

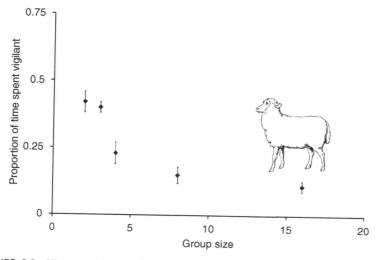

FIGURE 8.3 Vigilance in domesticated species. Foraging sheep decreased their allocation of time to vigilance in larger groups. Means and standard error bars are shown. *(Adapted from Michelena et al. (2012)).*

in larger groups. The confounding effect of density should be considered in future studies of vigilance conducted in confined spaces.

Domesticated species in large-scale, modern rearing conditions face little predation threats. But predation risk cannot so easily be ruled out in other cases. Chickens, for instance, were kept outdoors until recently where they were probably exposed to predation from cats (Burger, 1997). Wolves and other predators are also a threat to reindeer and free-ranging sheep and cattle. Despite domestication, the above species are still exposed to predation pressure, which may explain retention of the group-size effect on vigilance. Humans are also an obvious source of threats for domesticated species. Disturbances caused by humans are often considered equivalent to predation risk (Frid and Dill, 2002). The group-size effect on vigilance may persist nowadays in domesticated species partly as a response to threats from humans rather than predators.

Domesticated species may respond inappropriately to threats because of artificial selection for tameness. In cows, artificial selection for placid temperament has been linked to reduced vigilance and decreased reaction time to external disturbances (Flörcke et al., 2012). Individuals with such predispositions were more likely to lose calves to wolves in a free-range setting (Flörcke and Grandin, 2013). Such studies highlight the genetic underpinnings of vigilance and the need to better understand the current selection regime when predicting responses by domesticated species to natural enemies.

Domestication is an obvious way to alter the selection regime faced by a species. Similar changes in selection regime can also affect non-domesticated species. Some non-domesticated species, for instance, have been maintained for generations under conditions that limit encounters with natural predators. Species extirpated from the wild and kept in captivity have little contact with their natural predators. Reintroduction of such species in the wild allows us to examine how naïve animals respond to predation threats from their past. Reintroduction of Père David's deer in China is a good case in point. This species was extirpated from the wild more than 1000 years ago. The remaining deer were originally kept in captivity in China and sent later to zoos across the world. The reintroduced population in China originated from a small herd kept in Britain. Despite isolation from predators over hundreds of generations, Père David's deer still respond to the sights and sounds of their natural predators (Li et al., 2011a) and show the group-size effect on vigilance (Zheng et al., 2013). This species has retained a wariness of predators no longer a threat to current populations. However, other species have to relearn to be wary of potential predators (Griffin et al., 2000). Knowing how a species will react to threats from former foes and from new predators is critical to reintroduction programmes in the wild.

8.4.3   Vigilance on Islands and Predator-Free Habitats

Does vigilance occur in natural habitats where predators are absent or greatly reduced in numbers? Islands long isolated from predators offer a perfect test bed to explore this issue. It has long been known that lack of predation pressure on islands can lead to drastic changes in morphology and behaviour in prey animals (Quammen, 1996). On isolated islands, rattlesnakes have lost their rattle and normally shy birds are known to land on people's shoulders. In the absence of predation risk, any costly anti-predator adaptation, such as rattles, living in groups or vigilance, is expected to be eliminated by natural selection if it no longer serves a purpose (Lahti et al., 2009; Willis, 1972). Therefore, vigilance on islands isolated from predators should occur at low levels, if at all, and should not be associated with predation risk correlates like group size or distance to protective cover. These predictions have been tested recently in a handful of islands.

Controlling for a host of confounding factors, such as food density and time of day, vigilance failed to covary with group size in a bird species found on a predator-free island off Australia's Great Barrier Reef (Catterall et al., 1992). As we saw earlier, a lack of effect of group size on vigilance in birds is the exception rather than the rule, supporting the view that relaxed predation curtailed the need to adjust vigilance to group size in this bird species. This finding, however, may be unique to this species or to this particular island. Coincidence appears an unlikely explanation because similar findings have been reported for other species on different islands. For marsupial species in Australia, group size exerted less influence on vigilance on islands isolated from predators than on the predator-rich mainland (Blumstein and Daniel, 2005). In one telling example, isolation from predators over only 130 years was sufficient to reduce the effect of group size on vigilance in one species (Blumstein et al., 2004).

Lundy Island, off the United Kingdom, harbours a population of free-ranging Soay sheep that has been isolated from predators for decades now. Supporting the above findings in marsupials, vigilance in sheep was typically low and showed no relationship with group size (Hopewell et al., 2005). By contrast, on Rum Island, another UK island isolated from predators, vigilance in feral goats decreased with group size. But the researchers could not rule out competition both within and between species as a driver of vigilance in this species (Shi et al., 2010) (Fig. 8.4).

Vigilance is but one of many available anti-predator responses. Research indicates that loss of anti-predator responses in general following isolation from predators can be evident after only a few generations (Berger et al., 2001; Blumstein and Daniel, 2002; Blumstein et al., 2004; Stankowich and Coss, 2007). However, retention of ancestral traits in some species has been known despite thousands of years of isolation (Blumstein et al., 2000; Byers, 1997; Li et al., 2011a; Placyk and Burghardt, 2011). Is it possible for a trait like vigilance, which is typically viewed as costly, to persist under relaxed predation? Vigilance could persist if a prey species lost some but not all of its predators (Blumstein and Daniel, 2005; Coss, 1999). Even rare encounters with predators may be sufficient to maintain some anti-predator responses. Anti-predator vigilance could also be maintained if vigilance is useful in other contexts like

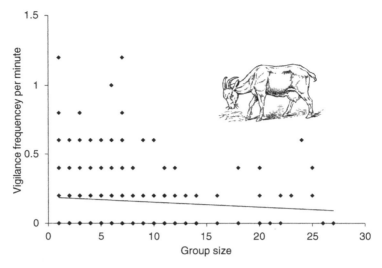

FIGURE 8.4 Vigilance with negligible predation risk. Scan frequency decreased in larger groups of feral goats on Rum Island, an island devoid of predators. The decline in vigilance in larger groups was explained by competition with conspecifics and other species. The trend line is shown. *(Adapted from Shi et al. (2010)).*

competition (Shi et al., 2010). The onus, therefore, is on researchers to document current predation pressure and the use of vigilance in other contexts.

The limited evidence gathered thus far indicates that relaxed predation pressure on isolated islands typically reduces the use of vigilance. Nevertheless, empirical studies of vigilance when predation risk is negligible face difficult challenges. Statistically, a lack of effect of group size on vigilance might reflect low statistical power rather than the absence of a biological effect. But even the situation where vigilance persists can be problematic. Should one invoke the ghost-of-predation past argument or contend that contacts with predators, while few and far between, have been sufficient to maintain vigilance? These remain open questions.

8.5 VIGILANCE AT NIGHT

Studies on vigilance have been mostly conducted during the day for practical reasons as night observations require expensive equipment and forces researchers to curtail sleep. However, the study of vigilance at night can be illuminating for several reasons. For prey animals active at night, maintaining vigilance at low light levels must certainly pose new challenges. Visual detection of predators is probably more difficult at night, which would select for increased vigilance at that time. On the other hand, vigilance may simply be curtailed in favour of other activities if the utility of vigilance at night is too low (Embar et al., 2011; Scheel, 1993). Not all predators are active at night and other sources of disturbances, such as hunters or passers-by, are often less frequent at night. Reduced predation pressure might select for lower night-time vigilance. Both group size and food availability can vary with time of day, and each of these factors, as we saw earlier, can influence vigilance levels. For all the above reasons, it is not easy to predict whether vigilance as a rule should increase or decrease at night. Another consideration is that light levels can vary tremendously at night depending on the amount of light reflected by the moon. Again, such changes in light regimes at night can influence anti-predator responses in prey animals (Kotler et al., 2010).

To investigate potential adjustments in vigilance at night, I contrasted vigilance in animal species that are active both day and night (Beauchamp, 2007). I noted whether the proportion of time allocated to vigilance was higher or lower at night than during the day, and assessed whether other factors, such as group size and the level of disturbances caused by people or predators, also varied concomitantly (Table 8.2). The findings revealed that vigilance was typically lower at night, with notable exceptions. In general, disturbances caused by people and predators tended to be less frequent at night. Group size and food availability at night also varied in many species (Beauchamp, 2007).

TABLE 8.2 Relative Trend in the Allocation of Time to Vigilance in Species that Forage both Day (D) and Night (N)

Species	Vigilance trend*	Other changes at night	References
Birds			
American avocet	D = N	Less disturbances at night	Kostecke and Smith (2003)
Dunlin	D = N	Changes in habitat; more food at night	Mouritsen (1992, 1994)
Trumpeter swan	D = N	Less disturbances at night	Henson and Cooper (1994)
White-faced whistling duck	D = N	Smaller groups at night	Petrie and Petrie (1998)
American flamingo	D < N	Less disturbances and smaller groups at night	Beauchamp and McNeil (2003)
Sanderling	D > N	Less disturbances and smaller groups at night	Burger and Gochfeld (1991)
Trumpeter swan	D > N	Less disturbances at night	McKelvey and Verbeek (1988)

(Continued)

TABLE 8.2 Relative Trend in the Allocation of Time to Vigilance in Species that Forage both Day (D) and Night (N) *(cont.)*

Species	Vigilance trend*	Other changes at night	References
Greylag goose	D > N	Less disturbances at night	Kahlert et al. (1996)
Mixed geese flocks	D > N		Mooij (1992)
Mallard	D > N		Jorde and Owen (1988)
Mallard	D > N		Jorde et al. (1984)
Mallard	D > N		Javurkova et al. (2011)
Mottled duck	D > N	Less disturbances at night	Paulus (1988)
American gadwall	D > N	Less disturbances at night	Paulus (1984)
White-headed duck	D > N	Less disturbances at night but no changes in group size	Green et al. (1999)
Mammals			
Eastern grey kangaroo	D = N		Clarke et al. (1989)
Agile wallaby	D = N	More disturbances at night; changes in habitat	Steer and Doody (2009)
Springbok	D < N	More disturbances at night	Bednekoff and Ritter (1994)
Tasmanian pademelon	D > N		Blumstein and Daniel (2003)
Quokka	D > N	No changes in group sizes or disturbance levels at night	Blumstein et al. (2001b)
Agile wallaby	D > N	Less disturbances at night	Stirrat (2004)
Eastern grey kangaroo	D > N	Less disturbances at night and changes in habitat	Southwell (1987)
African buffalo	D > N	More disturbances at night and changes in habitat	Prins (1996)
Mule deer	D > N	Less disturbances at night; changes in habitat and larger group sizes at night	McCullough (1993); McCullough et al. (1994)
European roe deer	D > N	Less disturbances at night	Sonnichsen et al. (2013)

*Entries are organized by type of responses

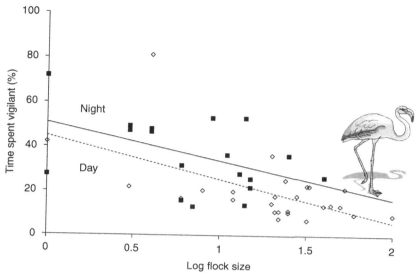

FIGURE 8.5   Vigilance at night. Wintering greater flamingos in coastal Venezuela spent more time vigilant at night (dark squares) than during the day (open diamonds), controlling for flock size, which suggests that the night was perceived as riskier. Trend lines are shown for each time of day. *(Adapted from Beauchamp and McNeil (2003)).*

Some species perceived the night as riskier. In American flamingos, vigilance was higher at night than during the day especially on moonless nights, suggesting that low light levels rather than the night *per se* were perceived as riskier (Beauchamp and McNeil, 2003) (Fig. 8.5). Low light levels at night were also perceived as riskier in Tammar wallabies (Biebouw and Blumstein, 2003).

It is not clear how vigilance can operate at low light levels. One possibility is that animals switch from visual to auditory vigilance, which might require longer scans to be effective. Auditory vigilance has been implicated recently in mule deer foraging at night (Lynch et al., 2015). The marker of vigilance, in this species, was a pause during rumination, which presumably decreased mastication noises. If vigilance at night is carried out principally with the ears, one would expect markers of auditory vigilance to go up.

As a rule, vigilance tended to decrease at night (Table 8.2). This finding supports the view that the night is perceived as safer than the day, which fits with the observation that disturbances caused by predators or people are less frequent at night. However, other adaptations to night foraging could also influence vigilance on their own. Agile wallabies, for example, were more likely to drink away from rivers at night to avoid lurking crocodiles (Steer and Doody, 2009). Such changes in habitat use, or in other factors like group size, must be taken into account when comparing day-time and night-time vigilance.

Low vigilance at night suggests that prey animals can use darkness as cover from predators. If this is the case, one would expect vigilance to increase when the night is more illuminated (Kotler et al., 2010; Nersesian et al., 2012). However, some species maintained less vigilance on brighter nights (Beauchamp and McNeil, 2003; Biebouw and Blumstein, 2003; Mella et al., 2014a). The picture might get clearer when the responses of more species to variation in light levels at night are reported.

Foraging at night might also affect collective detection and thus indirectly vigilance. At least for species that rely on visual cues, foraging at low light levels probably reduces the ability to detect alarm signals from neighbours. Although the efficiency of collective detection has never been investigated at night, as far as I know, some evidence suggests that the radius of interactions between neighbours could be lower at night. In the red-necked pademelon, a marsupial herbivore that forages primarily at night, neighbours only synchronized their vigilance when close to one another, suggesting that distant neighbours could not be easily monitored (Pays et al., 2009a). If collective detection is less efficient, dilution would represent the only factor available to reduce predation risk for species that forage in groups at night.

8.6 VIGILANCE IN MIXED-SPECIES GROUPS

Vigilance studies typically focus on species foraging in conspecific groups. Recent work on mixed-species groups indicates that species composition can be a determinant of vigilance just as well as the size of the group.

Mixed-species groups form when individuals from two or more species forage together with sufficient synchronization to suggest more than just a random assemblage of species. Mixed-species groups are fairly common in fish (Lukoschek and McCormick, 2002), birds (Sridhar et al., 2009), and mammals (Stensland et al., 2003). Some of the earliest investigations of vigilance actually focused on mixed-species flocks in birds (Bates, 1863; Belt, 1874). The interesting question with respect to vigilance is whether adding group members from other species brings more benefits than adding the same number of conspecifics.

Mingling with other species has been thought to improve the ability to detect predators (Altmann and Altmann, 1970; Thompson and Barnard, 1983). Some species may be better than others at detecting predators because they have sharper senses or special vigilance tactics (Heymann and Buchanan-Smith, 2000; Moore et al., 2013; Peres, 1993). By increasing the breadth of detection abilities at the group level, the formation of mixed-species groups could allow individuals to reduce their vigilance. Not surprisingly, species that are generally more alert or that respond to predators from a greater distance tend to attract other species to form mixed-species groups (Bshary and Noë, 1997; Byrkjedal and Kålås, 1983; Greig-Smith, 1981; Norris and Dohl, 1980; Semeniuk and Dill, 2006). Diana monkeys in the rainforests of Africa, for example, occupy greater heights when foraging and tend to respond to predators more quickly than other species. Red colobus monkeys exploit the sentinel qualities of Diana monkeys and experience a reduction in predation pressure from both ground and aerial predators when foraging close to Diana monkeys (Bshary and Noë, 1997). In mixed-species groups with Diana monkeys, red colobus monkeys looked sideways and down less often, suggesting a relaxation of vigilance.

Many studies have examined how prey animals adjust vigilance when foraging in mixed-species groups. In birds, a recent review of mixed-species flocks worldwide revealed that as a rule vigilance in mixed-species flocks tended to be lower than in single-species flocks (Sridhar et al., 2009). In mammals, the results have been mixed thus far with some studies documenting a decrease in vigilance in mixed-species groups compared to single-species groups (Cords, 1990; Hardie and Buchanan-Smith, 1997; Makenbach et al., 2013; Rasa, 1983; Scheel, 1993; Schmitt et al., 2014; Stojan-Dolar and Heymann, 2010b; Waterman and Roth, 2007; Wolters and Zuberbuhler, 2003), while others showed no changes at all (Bednekoff and Ritter, 1994; Pays et al., 2014; Porter and Garber, 2007; Treves, 1999), or even an increase (Chapman and Chapman, 1996; Creel et al., 2014; Stanford, 1998; Waterman and Roth, 2007). The latter cases suggest that other species can in fact pose a threat that requires extra monitoring (Barnard and Stephens, 1981; Popp, 1988a).

Comparing vigilance in single- and mixed-species groups is fraught with ambiguity. Mixed-species groups tend to be larger than single-species groups (Li et al., 2010; Stojan-Dolar and Heymann, 2010b). Lower vigilance in mixed-species groups might simply reflect the fact that such groups are larger. In the Diana-red colobus monkey association, researchers compared mixed-species and single-species groups of the same size to avoid this issue. The same approach was used in zebras (Schmitt et al., 2014) (Fig. 8.6). However, a control for group size is often lacking. Comparing vigilance in the two types of groups can also be difficult if mixed-species groups are the norm rather than the exception (Stojan-Dolar and Heymann, 2010b) or if mixed-species groups form at particular times of the year or only in special habitats with no equivalent for single-species groups. In primates, for instance, mixed-species and single-species groups often form at different forest strata where predation risk and foraging contingencies probably differ (Stojan-Dolar and Heymann, 2010b).

Other mechanisms could also explain why vigilance is lower in mixed-species groups. Joining a stronger species able to repel predators more effectively (Herzing and Johnson, 1997; Struhsaker, 1981) could reduce predation risk and drive vigilance down in the weaker species. Joining a species that is more attractive to predators might deflect predation pressure to other individuals allowing a reduction in vigilance (FitzGibbon, 1990; Mathis and Chivers, 2003). In line with this argument, zebras decreased their vigilance to a greater extent when associating with another prey species preferred by their shared predator, the lion, than with a non-preferred prey (Schmitt et al., 2014).

Despite all these potential benefits, the formation of mixed-species groups can also be costly in terms of vigilance. Odd-looking species in a rather

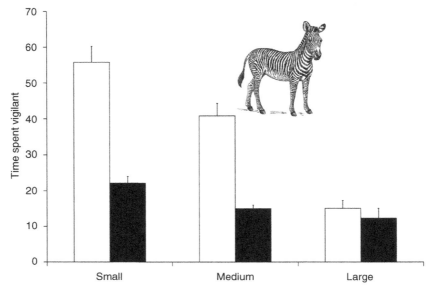

FIGURE 8.6 Vigilance in mixed-species groups. Zebras in mixed-species groups (black bars) maintained less vigilance (number of seconds spent vigilant per 3 min) than in single-species groups (white bars) of the same size when overall group size was small (2–9) or medium (10–30) but not large (31–75). Means and standard error bars are shown. *(Adapted from Schmitt et al. (2014)).*

uniform mixed-species group may be targeted preferentially by predators (Landeau and Terborgh, 1986). In this case, odd-looking species might be more vigilant than other species in the same group. Other species in a mixed-species group could also compete more effectively for resources or be more aggressive, which would force individuals of the less competitive species to invest more time in vigilance than would be the case had they joined conspecifics instead (Pays et al., 2014).

The pirate theory for the formation of mixed-species groups suggests that some species join groups to steal resources from others (Diamond, 1981). Species that are targeted more often by kleptoparasitic species would be expected to maintain more vigilance to reduce or avoid losses. However, theft in mixed-species groups might be tolerated because food-stealing species often act as sentinels for the group (Goodale and Kotagama, 2005; Ridley and Raihani, 2007). Their presence would allow individuals in the group to actually decrease their vigilance, a benefit that would compensate for occasionally losing resources to kleptoparasite species (Baigrie et al., 2014; Radford et al., 2011). The sentinel effect can function even for species that do not associate regularly in mixed-species groups. Dwarf mongooses, for instance, reduced their vigilance in the presence of a bird species that can detect predators more easily (Sharpe et al., 2010). Similarly, a solitary avian species decreased its vigilance in the presence of another species with a well-developed alarm system (Ridley et al., 2014).

8.7 CONCLUSIONS

Studying vigilance outside the box holds much promise for future work. Vigilance at night exposes species to low light levels. This can help us examine how vigilance is maintained when predator detection or mutual warning is impaired. Studying vigilance in mixed-species groups can address issues related to group composition, especially when participating species differ in their ability to detect predators. Studying vigilance when predation risk is negligible allows us to examine what happens when one of the major drivers of vigilance becomes less potent.

Vigilance can also be useful as a tool to assess the sensitivity of animals to human disturbances. In view of the varied responses to different types of threats and the equally varied responses by different species to the same threats, future work should aim to explain why such variability occurs amongst species in response to disturbances. Sensitivity to disturbances suggests that vigilance could also be used as a tool to assess animal welfare (Welp et al., 2004). High vigilance of the costly type could be interpreted as a sign that a situation is particularly threatening.

Work in these relatively new areas of research will expand the scope of vigilance studies and shed new light on old issues. Studying vigilance outside the box can be viewed as an alternative to the same old investigations focusing on traditional factors like group size and sex.

Conclusions

WHAT HAVE WE LEARNED?

Scientific interest in the study of animal vigilance dates back to the end of the nineteenth century. British naturalists laid the foundation for this research by asking how animals could benefit from living in groups. It became clear quite early that vigilance can help detect predation threats more rapidly. Threats are less likely to remain undetected for long, especially when many eyes and ears command all approaches. Fright reactions following the detection of predation threats can also pass quickly from detectors to non-detectors allowing everyone to flee rapidly.

The role of vigilance in the detection of predation threats was established early in the literature, and yet it took nearly 100 years for researchers to model the detection process and realize that animals in groups could actually reduce their investment in vigilance at no increased risk to themselves. Early models established that in groups the detection of threats should occur earlier and vigilance should be lower. These two predictions have been the cornerstone of empirical vigilance projects for the past 40 years.

Predators are not the only target of vigilance. Research with geese, pigeons, and primates recognized rivals from the same species as alternative targets of vigilance. Because social vigilance is predicted to increase with the frequency of interactions amongst individuals, this tendency can in fact counteract the expected decrease in anti-predator vigilance in larger groups.

Vigilance research concentrates on functional questions: in particular, how does vigilance aimed at predators or rivals help individuals increase their reproductive success or survival, and thus ultimately their fitness? Questions addressing developmental, mechanistic or evolutionary issues have comparatively attracted much less attention. The recent surge of interest in non-functional explanations of behaviour bodes well for future research on these questions.

Any environmental or internal factor that influences social or predation risk is likely to affect vigilance. Group size is certainly the environmental factor that has received the most attention, but many other drivers of vigilance have been discussed, including energy reserves, distance to cover, and weather. Research on the drivers of vigilance has received considerable attention from animal vigilance students.

Animal Vigilance. http://dx.doi.org/10.1016/B978-0-12-801983-2.00009-7

Theoretical models have been an integral part of vigilance research from the beginning, and continue to this day to provide testable predictions and explanations for empirical patterns of vigilance. Many assumptions made in early models have been questioned over the years. In particular, the temporal organization of vigilance bouts often fails to fit the sort of randomness embodied in these models. In addition, the assumption of independent vigilance amongst group members has been challenged in many species. Such challenges should be viewed as an opportunity to enrich the theoretical framework behind vigilance research as it moves towards Vigilance 2.0.

Vigilance research attracts students of animal behaviour and behavioural ecology. As such, it has inherited certain attributes by default. Researchers thus tend to adopt a functional perspective on behaviour and focus on animals in their natural habitats. Research on vigilance outside this box may turn out to be a welcome addition. The study of vigilance in humans or domesticated animals is fairly new but promises to address issues regarding the target of vigilance in species facing little predation pressure. Species foraging on isolated islands and at night often experience relaxed predation pressure as well, which provides us with an opportunity to examine what happens to vigilance where one major target of vigilance is missing. Vigilance can also be used to document responses to human disturbances or to address welfare issues.

WHERE DO WE GO FROM HERE?

I end this book by articulating what are for me the most pressing issues for the future of vigilance research. In no particular order, I identified the following research questions.

1. Identification of the markers of vigilance

   It has become clear over the years that the postural definition of vigilance is wanting. There are many reasons why an animal may raise its head only one of them has to do with vigilance. In addition, it is not always easy to ascertain the target of vigilance from a posture. This is particularly troubling because many drivers affect anti-predator and social vigilance in drastically different ways. The other troubling aspect of the postural definition is that it largely restricts vigilance research to birds and mammals.

   The internal state of vigilance is not necessarily obvious from external signs. Unless we can probe the brains of our study species in real time, this shall remain so for most research projects. One solution is to skirt the definition of vigilance altogether and simply focus on its consequences. If there is a trade-off between foraging and vigilance, an animal that invests more time on vigilance will experience a reduction in food intake rate. The amount of food ingested during a fixed period of time can thus be used as an indirect measure of the investment in vigilance (Brown, 1988). This approach, in fact, does not even require direct observation if the amount of food eaten by an animal in a patch can be evaluated after foraging. Such an approach works best when measurements are made over a short time period during which no other competing activities take place. This is important because animals may alter time spent resting or grooming to maintain the same feeding rate despite changes in predation risk. This approach also works best for solitary foragers. Measuring individual food intake in a group may be difficult and group foraging introduces competition issues. In a group, this approach also conflates anti-predator and social vigilance.

Vigilance also has another consequence, namely, the probability of detecting a threat. Vigilance could thus also be inferred indirectly from the ability to detect threats. This approach requires experimental exposure to a standardized threat and can only work if exposed animals show an obvious reaction to the threat. Many factors other than vigilance can influence detection, including fatigue and environmental obstacles, and these factors need to be taken into account.

An obvious advantage of these indirect approaches to measuring vigilance is that they apply to any type of animals exposed to risk. However, such approaches require interventions by researchers to measure food intake or to simulate predation threats. If the goal is to investigate animals in a natural setting instead, it becomes essential to define vigilance not by its consequences but by its observable attributes. In this situation, we need to develop observable markers of the state of vigilance and validate their use.

Markers may involve internal or external features. Promising internal markers include changes in heart rate or hormone levels associated with fear. External markers, such as changes in postures, do not require invasive measurements but may be less sensitive. When a putative marker of vigilance has been identified, it should be validated by measuring the ability to detect threats. For instance, if raising the head to scan the surroundings does not increase the ability to detect a standardized threat, it cannot be used as a marker of vigilance.

Knowledge of the sensory ability of our study species will also prove most helpful in defining markers of vigilance (Stevens, 2013). Such knowledge can also help us determine the target of vigilance. As an example, if the visual field of a species is relatively narrow, the target of vigilance could be deduced from gaze direction. Solving all the previous issues will broaden the scope of vigilance research and provide us with a firmer foundation to test predictions from theoretical models.

2. The group-size effect on vigilance

Despite hundreds of paper on the group-size effect on vigilance, several issues remain unresolved. A meta-analysis of the group-size effect on vigilance is lacking in mammals. Therefore, it is not clear whether group size in mammals is as good a predictor of vigilance as it is in birds. Mammals are typically larger than birds, live longer and tend to be more sedentary. Different life history trajectories may influence the relationship between vigilance and group size. Extension of vigilance research to species other than birds and mammals will also prove most helpful to determine the robustness of the effects documented thus far.

In terms of mechanisms, little is known about the ability of animal species to estimate group size even though evolutionarily stable vigilance has always been expressed as a function of group size. If animals cannot count beyond two, say, we should not expect fine scale adjustments in vigilance to group size. It may also turn out that individuals in a group only interact with close neighbours rather than with everyone in sight, in which case the standard definition of group size will not work.

In terms of development, we still do not know if animals need to learn that larger groups are safer. Knowing how and when experiences come into play will help us to understand age-related changes in vigilance and the ability to adjust vigilance to group size. How animals reach the evolutionarily stable level of vigilance is left unspecified by most models of vigilance. Do they watch one another and negotiate the outcome or use fixed strategies to determine the proper level of vigilance?

Plots of the relationship between vigilance and group size have revealed substantial heterogeneity in the scatter, especially at smaller group sizes (Beauchamp, 2013b). One possibility to explain this wide scatter is that the relationship between vigilance and group size is shallower or steeper depending on the context faced by animals. For instance, vigilance may decrease more slowly with group size in juveniles than in adults or at the edges of the group than in the centre. A challenge for future research is to document the sources of heterogeneity in the group-size effect on vigilance.

Nearly all empirical studies thus far have investigated the qualitative relationship between vigilance and group size, namely, whether vigilance decreases or not with group size, rather than the quantitative predictions derived from vigilance models. Claiming success if vigilance decreases with group size can be misleading if observed vigilance levels are over- or underestimated by a wide margin. It is important to understand deviation from both qualitative and quantitative predictions. Testing quantitative predictions is conceptually simple: fill in the blanks for input variables in a model and calculate the predicted level of vigilance for a given group size. The devil is in getting the information needed to populate input parameters of the models. The next step is obviously to gather the information needed to test quantitative predictions.

3. Theoretical models

Vigilance models specify how much time should be allocated to vigilance to maximize fitness, but remain silent on how animals practically achieve the expected level of vigilance. Animals can alter both the duration and the frequency of scanning to achieve a desired level of vigilance. The different means of achieving the same level of vigilance are not necessarily equivalent. Irregular long scans rather than frequent short ones, for instance, may not be useful if predators can move in quickly to attack prey animals. We need models that can specify the frequency and duration of scans in addition to the overall amount of time spent scanning. To achieve this goal will require a better understanding of the relationship between the duration of scan and interscan intervals and fitness. For instance, the ability to detect a threat should increase with scan duration but probably not in a linear fashion, and should vary with the sensory ability of a species. This type of modelling will allow us to incorporate mechanistic considerations into the more traditional functional framework.

4. Drivers of vigilance

Drivers of vigilance influence the risk perceived by individuals. Except for group size, no meta-analyses are available to assess the evidence that internal or external factors influence vigilance. In this book, I simply made a list of studies supporting or not qualitative predictions for each putative driver. A formal meta-analysis is required to determine how much variation in vigilance is explained by each driver. For most drivers, I detected many cases that poorly fitted the predicted patterns. The next step is to understand why the general pattern does not hold in every case.

Answering all of the previous questions will keep students of animal vigilance busy for the next 20 years. As the saying goes, eternal vigilance is of the essence when it comes to monitoring human affairs. I hope that this book will show how to keep vigilance research a ceaseless topic of interest to animal behaviour students.

References

Afshar, M., Giraldeau, L.-A., 2014. A unified modelling approach for producer–scrounger games in complex ecological conditions. Anim. Behav. 96, 167–176.

Alados, C.L., 1985. An analysis of vigilance in the Spanish ibex (Capra pyrenaica). Z. Tierpsychol. 68, 58–64.

Alberts, S.C., 1994. Vigilance in young baboons: effects of habitat, age, sex and maternal rank on glance rate. Anim. Behav. 47, 749–755.

Alcock, J., 1975. Animal Behavior: An Evolutionary Approach. Sinauer Associates, Sunderland.

Ale, S.B., Brown, J.S., 2007. The contingencies of group size and vigilance. Evol. Ecol. Res. 9, 1263–1276.

Allee, W.C., 1931. Animal Aggregations: A Study in General Sociology. University of Chicago Press, Chicago.

Allee, W.C., 1932. Animal Life and Social Growth. William and Wilkins, Baltimore.

Allen, W.E., 1920. Behavior of loon and sardines. Ecology 1, 309–310.

Alonso, J.A., Alonso, J.C., 1993. Age-related differences in time budgets and parental care in wintering common cranes. Auk 110, 78–88.

Altmann, M., 1951. Social behaviour of elk, Cervus canadensis nelsoni, in the Jackson Hole area of Wyoming. Behaviour 4, 116–142.

Altmann, M., 1958. The flight distance in free ranging big mammals. J. Wildl. Manage. 22, 207–209.

Altmann, S.A., Altmann, J., 1970. Baboon Ecology: African Field Research. University of Chicago Press, Chicago.

Amano, T., Ushiyama, K., Fujita, G., Higuchi, H., 2006. Costs and benefits of flocking in foraging white-fronted geese (Anser albifrons): effects of resource depletion. J. Zool. 269, 111–115.

Amlaner, C.J., McFarland, D.J., 1981. Sleep in the herring gull (Larus argentatus). Anim. Behav. 29, 551–556.

Anderies, J.M., Katti, M., Schochat, E., 2007. Living in the city: resource availability, predation, and bird population dynamics in urban areas. J. Theor. Biol. 247, 36–49.

Andrew, R.J., 1978. Increased persistence of attention produced by testosterone, and its implications for the study of sexual behaviour. In: Hutchinson, J.B. (Ed.), Sexual Behaviour. John Wiley, Chichester, pp. 255–275.

Andrews, M.I., Naik, R.M., 1970. The biology of the jungle babbler. Pavo 8, 1–34.

Apfelbach, R., Blanchard, C.D., Blanchard, R.J., Hayes, R.A., McGregor, I.S., 2005. The effects of predator odors in mammalian prey species: a review of field and laboratory studies. Neurosci. Biobehav. Rev. 29, 1123–1144.

Arenz, C.L., Leger, D.W., 1997a. The antipredator vigilance of adult and juvenile thirteen-lined ground squirrels (Sciuridae: Spermophilus tridecemlineatus): visual obstruction and simulated hawk attacks. Ethology 103, 945–953.

Animal Vigilance. http://dx.doi.org/10.1016/B978-0-12-801983-2.00010-3

Arenz, C.L., Leger, D.W., 1997b. Artificial visual obstruction, antipredator vigilance, and predator detection in the thirteen-lined ground squirrel (*Spermophilus tridecemlineatus*). Behaviour 134, 1101–1114.

Arenz, C.L., Leger, D.W., 1999. Thirteen-lined ground squirrel (Sciuridae: *Spermophilus tridecemlineatus*) antipredator vigilance: monitoring the sky for aerial predators. Ethology 105, 807–816.

Arenz, C.L., Leger, D.W., 2000. Antipredator vigilance of juvenile and adult thirteen-lined ground squirrels and the role of nutritional need. Anim. Behav. 59, 535–541.

Armitage, K.B., Corona, M.C., 1994. Time and wariness in yellow-bellied marmots. J. Mt. Ecol. 2, 1–8.

Arnqvist, G., Wooster, D., 1995. Meta-analysis: synthesizing research findings in ecology and evolution. Trends Ecol. Evol. 10, 236–240.

Artiss, T., Hochachka, W.M., Martin, K., 1999. Female foraging and male vigilance in white-tailed ptarmigan (*Lagopus leucurus*): opportunism or behavioural coordination? Behav. Ecol. Sociobiol. 46, 429–434.

Artiss, T., Martin, K., 1995. Male vigilance in white-tailed ptarmigans, *Lagopus leucurus*: mate guarding or predator detection? Anim. Behav. 49, 1249–1258.

Austin, J., 1990. Comparison of activities within families and pairs of wintering Canada geese. Wilson Bull. 102, 536–542.

Austin, N.P., Rogers, L.J., 2012. Limb preferences and lateralization of aggression, reactivity and vigilance in feral horses, *Equus caballus*. Anim. Behav. 83, 239–247.

Austin, N.P., Rogers, L.J., 2014. Lateralization of agonistic and vigilance responses in Przewalski horses (*Equus przewalskii*). Appl. Anim. Behav. Sci. 151, 43–50.

Avilés, J.M., 2003. Time budget and habitat use of the common crane wintering in dehesas of southwestern Spain. Can. J. Zool. 81, 1233–1238.

Avilés, J.M., Bednekoff, P.A., 2007. How do vigilance and feeding by common cranes *Grus grus* depend on age, habitat, and flock size? J. Avian Biol. 38, 690–697.

Bachman, G.C., 1993. The effect of body condition on the trade-off between vigilance and foraging in Belding's ground squirrels. Anim. Behav. 46, 233–244.

Badzinski, S.S., 2005. Social influences on tundra swan activities during migration. Waterbirds 28, 316–325.

Bahr, D.B., Bekoff, M., 1999. Predicting flock vigilance from simple passerine interactions: modelling with cellular automata. Anim. Behav. 58, 831–839.

Baigrie, B.D., Thompson, A.M., Flower, T.P., 2014. Interspecific signalling between mutualists: food-thieving drongos use a cooperative sentinel call to manipulate foraging partners. Proc. R. Soc. Lond. B 281, 41232.

Baker, D.J., Stillman, R.A., Smart, S.L., Bullock, J.M., Norris, K.J., 2011. Are the costs of routine vigilance avoided by granivorous foragers? Funct. Ecol. 25, 617–627.

Balda, R.P., Bateman, G.C., 1971. Flocking and the annual cycle of the pinon jay, *Gymnorhinus cyanocephalus*. Condor 73, 287–302.

Baldellou, M., Henzi, S.P., 1992. Vigilance, predator detection and the presence of supernumerary males in vervet monkey troops. Anim. Behav. 43, 451–461.

Ball, E.E., 1977. Fine structure of the compound eyes of the midwater amphipod *Phronima* in relation to behavior and habitat. Tissue Cell 9, 521–536.

Ball, N., Shaffery, J.P., Amlaner, C.J., 1984. Sleeping gulls and predator avoidance. Anim. Behav. 34, 1253–1254.

Barber, I., Hoare, D., Krause, J., 2000. Effects of parasites on fish behaviour: a review and evolutionary perspective. Rev. Fish Biol. Fisher. 10, 131–165.

Barber, J.R., Crooks, K.R., Fristrup, K.M., 2010. The costs of chronic noise exposure for terrestrial organisms. Trends Ecol. Evol. 25, 180–189.

Barbosa, A., 2002. Does vigilance always covary negatively with group size? Effects of foraging strategy. Acta Ethol. 5, 51–55.

Barnard, C.J., 1980. Flock feeding and time budgets in the house sparrow (*Passer domesticus* L.). Anim. Behav. 28, 295–309.

Barnard, C.J., Sibly, R.M., 1981. Producers and scroungers: a general model and its application to captive flocks of house sparrows. Anim. Behav. 29, 543–550.

Barnard, C.J., Stephens, H., 1981. Prey selection by lapwings in lapwing–gull associations. Behaviour 77, 1–22.

Barnard, C.J., Stephens, H., 1983. Costs and benefits of single and mixed-species flocking in fieldfares (*Turdus pilaris*) and redwings (*T. iliacus*). Behaviour 84, 91–123.

Barnard, C.J., Thompson, D.B.A., 1985. Gulls and Plovers: The Ecology and Behaviour of Mixed-Species Feeding Groups. Croom Helm, London.

Barros, M., Alencar, C., Silva, M.A.d.S., Tomaz, C., 2008. Changes in experimental conditions alter anti-predator vigilance and sequence predictability in captive marmosets. Behav. Proc. 77, 351–356.

Barta, Z., Flynn, R., Giraldeau, L.-A., 1997. Geometry for a selfish foraging group: a genetic algorithm approach. Proc. R. Soc. Lond. B 264, 1233–1238.

Barta, Z., Liker, A., Monus, F., 2004. The effects of predation risk on the use of social foraging tactics. Anim. Behav. 67, 301–308.

Basile, M., Boivin, S., Boutin, A., Blois-Heulin, C., Hausberger, M., Lemasson, A., 2009. Socially dependent auditory laterality in domestic horses (*Equus caballus*). Anim. Cogn. 12, 611–619.

Bates, H.W., 1863. The Naturalist on the River Amazons. Murray Press, London.

Bateson, P.P.G., 1978. How does behavior develop? In: Bateson, P.P.G., Klopfer, P.H. (Eds.), Social Behavior. Springer, US, pp. 55–66.

Beale, C.M., Monaghan, P., 2004a. Behavioural responses to human disturbance: a matter of choice? Anim. Behav. 68, 1065–1069.

Beale, C.M., Monaghan, P., 2004b. Human disturbance: people as predation-free predators? J. Appl. Ecol. 41, 335–343.

Beauchamp, G., 2001. Should vigilance always decrease with group size? Behav. Ecol. Sociobiol. 51, 47–52.

Beauchamp, G., 2006. Non-random patterns of vigilance in flocks of the greater flamingo, *Phoenicopterus ruber ruber*. Anim. Behav. 71, 593–598.

Beauchamp, G., 2007. Exploring the role of vision in social foraging: what happens to group size, vigilance, spacing, aggression and habitat use in birds and mammals that forage at night? Biol. Rev. 82, 511–525.

Beauchamp, G., 2008a. A spatial model of producing and scrounging. Anim. Behav. 76, 1935–1942.

Beauchamp, G., 2008b. What is the magnitude of the group-size effect on vigilance? Behav. Ecol. 19, 1361–1368.

Beauchamp, G., 2009. Sleeping gulls monitor the vigilance behaviour of their neighbours. Biol. Lett. 5, 9–11.

Beauchamp, G., 2010a. A comparative analysis of vigilance in birds. Evol. Ecol. 24, 1267–1276.

Beauchamp, G., 2010b. Determinants of false alarms in staging flocks of semipalmated sandpipers. Behav. Ecol. 21, 584–587.

Beauchamp, G., 2010c. Relationship between distance to cover, vigilance and group size in staging flocks of semipalmated sandpipers. Ethology 116, 645–652.

Beauchamp, G., 2011. Collective waves of sleep in gulls (*Larus* spp.). Ethology 117, 326–331.

Beauchamp, G., 2013a. Foraging success in a wild species of bird varies depending on which eye is used for anti-predator vigilance. Laterality 18, 194–202.

Beauchamp, G., 2013b. Is the magnitude of the group-size effect on vigilance underestimated? Anim. Behav. 85, 281–285.

Beauchamp, G., 2014a. Antipredator vigilance decreases with food density in staging flocks of semipalmated sandpipers (*Calidris pusilla*). Can. J. Zool. 92, 785–788.

Beauchamp, G., 2014b. Social Predation: How Group Living Benefits Predators and Prey. Academic Press, New York.

Beauchamp, G., Alexander, P., Jovani, R., 2012. Consistent waves of collective vigilance in groups using public information about predation risk. Behav. Ecol. 23, 368–374.

Beauchamp, G., Livoreil, B., 1997. The effect of group size on vigilance and feeding rate in spice finches (*Lonchura punctulata*). Can. J. Zool. 75, 1526–1531.

Beauchamp, G., McNeil, R., 2003. Vigilance in greater flamingos foraging at night. Ethology 109, 511–520.

Beauchamp, G., McNeil, R., 2004. Levels of vigilance track changes in flock size in the greater flamingo (*Phoenicopterus ruber ruber*). Ornitol. Neotrop. 15, 407–411.

Beauchamp, G., Ruxton, G.D., 2003. Changes in vigilance with group size under scramble competition. Am. Nat. 161, 672–675.

Beauchamp, G., Ruxton, G.D., 2007. Dilution games: use of protective cover can cause a reduction in vigilance for prey in groups. Behav. Ecol. 18, 1040–1044.

Beauchamp, G., Ruxton, G.D., 2008. Disentangling risk dilution and collective detection in the antipredator vigilance of semipalmated sandpipers in flocks. Anim. Behav. 75, 1837–1842.

Beauchamp, G., Ruxton, G.D., 2011. A reassessment of the predation risk allocation hypothesis: a comment on Lima and Bednekoff. Am. Nat. 177, 143–146.

Beauchamp, G., Ruxton, G.D., 2012a. Changes in anti-predator vigilance over time caused by a war of attrition between predator and prey. Behav. Ecol. 23, 368–374.

Beauchamp, G., Ruxton, G.D., 2012b. Vigilance decreases with time at loafing sites in gulls (*Larus* spp.). Ethology 118, 733–739.

Bednekoff, P.A., 1997. Mutualism among safe, selfish sentinels: a dynamic game. Am. Nat. 150, 373–390.

Bednekoff, P.A., Blumstein, D.T., 2009. Peripheral obstructions influence marmot vigilance: integrating observational and experimental results. Behav. Ecol. 20, 1111–1117.

Bednekoff, P.A., Bowman, R., Woolfenden, G.E., 2008. Do conversational gutturals help Florida scrub-jays coordinate their sentinel behavior? Ethology 114, 313–317.

Bednekoff, P.A., Lima, S.L., 1998a. Randomness, chaos and confusion in the study of antipredator vigilance. Trends Ecol. Evol. 13, 284–287.

Bednekoff, P.A., Lima, S.L., 1998b. Re-examining safety in numbers: interactions between risk dilution and collective detection depend upon predator targeting behaviour. Proc. R. Soc. Lond. B 265, 2021–2026.

Bednekoff, P.A., Lima, S.L., 2002. Why are scanning patterns so variable? An overlooked question in the study of anti-predator vigilance. J. Avian Biol. 33, 143–149.

Bednekoff, P.A., Lima, S.L., 2004. Risk allocation and competition in foraging groups: reversed effects of competition if group size varies under risk of predation. Proc. R. Soc. Lond. B 271, 1491–1496.

Bednekoff, P.A., Lima, S.L., 2005. Testing for peripheral vigilance: do birds value what they see when not overtly vigilant? Anim. Behav. 69, 1165–1171.

Bednekoff, P.A., Ritter, R., 1994. Vigilance in Nxai pan springbok, *Antidorcas marsupialis*. Behaviour 129, 1–11.

Bednekoff, P.A., Woolfenden, G.E., 2003. Florida scrub-jays (*Aphelocoma coerulescens*) are sentinels more when well-fed (even with no kin nearby). Ethology 109, 895–903.

Bednekoff, P.A., Woolfenden, G.E., 2006. Florida scrub-jays compensate for the sentinel behavior of flockmates. Ethology 112, 796–800.

Bekoff, M., 1995. Vigilance, flock size, and flock geometry: information gathering by western evening grosbeaks (Aves: *Fringillidae*). Ethology 99, 150–161.

Bekoff, M., 1996. Cognitive ethology, vigilance, information gathering, and representation: who might know what and why? Behav. Proc. 35, 225–237.

Bélanger, L., Bédard, J., 1990. Energetic cost of man-induced disturbance to staging snow geese. J. Wildl. Manage. 54, 36–41.

Bélanger, L., Bédard, J., 1992. Flock composition and foraging behaviour of greater snow geese (*Chen caerulescens atlantica*). Can. J. Zool. 70, 2410–2415.

Bell, M.B.V., Radford, A.N., Rose, R., Wade, H.M., Ridley, A.R., 2009. The value of constant surveillance in a risky environment. Proc. R. Soc. Lond. B 276, 2997–3005.

Belt, T.W., 1874. The Naturalist in Nicaragua. Murray Press, London.

Benhaiem, S., Delon, M., Lourtet, B., Cargnelutti, B., Aulagnier, S., Hewison, A.J.M., Morellet, N., Verheyden, H., 2008. Hunting increases vigilance levels in roe deer and modifies feeding site selection. Anim. Behav. 76, 611–618.

Benoist, S., Garel, M., Cugnasse, J.M., Blanchard, P., 2013. Human disturbances, habitat characteristics and social environment generate sex-specific responses in vigilance of Mediterranean mouflon. Plos One 8, e82960.

Beránková, J., Veselý, P., Sýkorová, J., Fuchs, R., 2014. The role of key features in predator recognition by untrained birds. Anim. Cogn. 17, 963–971.

Berger, J., 1978. Group size, foraging, and antipredator ploys: an analysis of bighorn sheep decisions. Behav. Ecol. Sociobiol. 4, 91–99.

Berger, J., 1991. Pregnancy incentives, predation constraints and habitat shifts: experimental and field evidence for wild bighorn sheep. Anim. Behav. 41, 61–77.

Berger, J., Cunningham, C., 1988. Size-related effects on search times in North American grassland female ungulates. Ecology 69, 177–183.

Berger, J., Cunningham, C., 1995. Predation, sensitivity and sex: why female black rhinoceroses outlive males. Behav. Ecol. 6, 57–64.

Berger, J., Daneke, J., Johnson, J., Berwick, S.H., 1983. Pronghorn foraging economy and predator avoidance in a desert ecosystem: implications for the conservation of large mammalian herbivores. Biol. Conserv. 25, 193–208.

Berger, J., Swenson, J.E., Persson, I.-L., 2001. Recolonizing carnivores and naive prey: conservation lessons from Pleistocene extinctions. Science 291, 1036–1039.

Bergvall, U.A., Schäpers, A., Kjellander, P., Weiss, A., 2011. Personality and foraging decisions in fallow deer *Dama dama*. Anim. Behav. 81, 101–112.

Bertram, B.C.R., 1978. Living in groups: predator and prey. In: Krebs, J.R., Davies, N.B. (Eds.), Behavioural Ecology. Blackwell, Oxford, pp. 64–96.

Bertram, B.C.R., 1980. Vigilance and group size in ostriches. Anim. Behav. 28, 278–286.

Betts, B.J., 1976. Behaviour in a population of Columbian ground squirrels, *Spermophilus columbianus columbianus*. Anim. Behav. 24, 652–680.

Beveridge, F.M., Deag, J.M., 1987. The effects of sex, temperature and companions on looking up and feeding in single and mixed species flocks of house sparrows (*Passer domesticus*), chaffinches (*Fringilla coelebs*) and starlings (*Sturnus vulgaris*). Behaviour 100, 303–320.

Biebouw, K., Blumstein, D.T., 2003. Tammar wallabies (*Macropus eugenii*) associate safety with higher levels of nocturnal illumination. Ethol. Ecol. Evol. 15, 159–172.

Biegler, R., McGregor, A., Krebs, J.R., Healy, S.D., 2001. A larger hippocampus is associated with longer-lasting spatial memory. Proc. Natl. Acad. Sci. 98, 6941–6944.

Birke, L., 2002. Effects of browse, human visitors and noise on the behaviour of captive orangutans. Anim. Welf. 11, 189–202.

Black, J.M., Carbone, C., Wells, R.L., Owen, M., 1992. Foraging dynamics in goose flocks: the cost of living on the edge. Anim. Behav. 44, 41–50.

Black, J.M., Owen, M., 1989. Parent–offspring relationships in wintering barnacle geese. Anim. Behav. 37, 187–198.

Blackman, N., Proschau, F., 1957. Optimum search for objects having unknown arrival times. Oper. Res. 7, 625–638.

Blanchard, P., Fritz, H., 2007. Induced or routine vigilance while foraging. Oikos 116, 1603–1608.

Blanchard, P., Sabatier, R., Fritz, H., 2008. Within-group spatial position and vigilance: a role also for competition? The case of impalas (*Aepyceros melampus*) with a controlled food supply. Behav. Ecol. Sociobiol. 62, 1863–1868.

Blumstein, D.T., 1996. How much does social group size influence golden marmot vigilance? Behaviour 133, 1133–1151.

Blumstein, D.T., Barrow, L., Luterra, M., 2008. Olfactory predator discrimination in yellow-bellied marmots. Ethology 114, 1135–1143.

Blumstein, D.T., Daniel, J.C., 2002. Isolation from mammalian predators differentially affects two congeners. Behav. Ecol. 13, 657–663.

Blumstein, D.T., Daniel, J.C., 2003. Foraging behavior of three Tasmanian macropodid marsupials in response to present and historical predation threat. Ecography 26, 585–594.

Blumstein, D.T., Daniel, J.C., 2005. The loss of anti-predator behaviour following isolation on islands. Proc. R. Soc. Lond. B 272, 1663–1668.

Blumstein, D.T., Daniel, J.C., Ardron, J.G., Evans, C.S., 2002. Does feeding competition influence tammar wallaby time allocation? Ethology 108, 937–945.

Blumstein, D.T., Daniel, J.C., Evans, C.S., 2001a. Yellow-footed rock-wallaby group size effects reflect a trade-off. Ethology 107, 655–664.

Blumstein, D.T., Daniel, J.C., Griffin, A.S., Evans, C.S., 2000. Insular tammar wallabies (*Macropus eugenii*) respond to visual but not acoustic cues from predators. Behav. Ecol. 11, 528–535.

Blumstein, D.T., Daniel, J.C., McLean, I.G., 2001b. Group size effects in quokkas. Aust. J. Zool. 49, 641–649.

Blumstein, D.T., Daniel, J.C., Sims, R.A., 2003. Group size but not distance to cover influences agile wallaby (*Macropus agilis*) time allocation. J. Mammal. 84, 197–204.

Blumstein, D.T., Daniel, J.C., Springett, B.P., 2004. A test of the multi-predator hypothesis: rapid loss of antipredator behavior after 130 years of isolation. Ethology 110, 919–934.

Blumstein, D.T., Evans, C.S., Daniel, J.C., 1999. An experimental study of behavioural group size effects in tammar wallabies, *Macropus eugenii*. Anim. Behav. 58, 351–360.

Blumstein, D.T., Lea, A.J., Olson, L.E., Martin, J.G.A., 2010. Heritability of anti-predatory traits: vigilance and locomotor performance in marmots. J. Evol. Biol. 23, 879–887.

Bohlin, T., Johnsson, J.I., 2004. A model on foraging activity and group size: can the relative contribution of predation risk dilution and competition be evaluated experimentally? Anim. Behav. 68, F1–F5.

Bohorquez-Herrera, J., Kawano, S.M., Domenici, P., 2013. Foraging behavior delays mechanically-stimulated escape responses in fish. Integr. Comp. Biol. 53, 780–786.

Boinski, S., Kauffman, L., Westoll, A., Stickler, C.M., Cropp, S., Ehmke, E., 2003. Are vigilance, risk from avian predators and group size consequences of habitat structure? A comparison of three species of squirrel monkey (*Saimiri oerstedii*, *S. boliviensis*, and *S. sciureus*). Behaviour 140, 1421–1467.

Boland, C.R.J., 2003. An experimental test of predator detection rates using groups of free-living emus. Ethology 109, 209–222.

Bonati, B., Csermely, D., 2010. Complementary lateralisation in the exploratory and predatory behaviour of the common wall lizard (*Podarcis muralis*). Laterality 11, 1–9.

Borkowski, J.J., White, P.J., Garrott, R.A., Davis, T., Hardy, A.R., Reinhart, D.J., 2006. Behavioral responses of bison and elk in Yellowstone to snowmobiles and snow coaches. Ecol. Appl. 16, 1911–1925.

Boukhriss, J., Selmi, S., Bechet, A., Nouira, S., 2007. Vigilance in greater flamingos wintering in southern Tunisia: age-dependent flock size effect. Ethology 113, 377–385.

Boyle, S.A., Samson, F.B., 1985. Effects of nonconsumptive recreation on wildlife: a review. Wildl. Soc. B 13, 110–116.

Boysen, A.F., Lima, S.L., Bakken, G.S., 2001. Does the thermal environment influence vigilance behavior in dark-eyed juncos (*Junco hyemalis*)? An approach using standard operative temperature. J. Therm. Biol. 26, 605–612.

Brick, O., 1998. Fighting behaviour, vigilance and predation risk in the cichlid fish, *Namacara anomala*. Anim. Behav. 56, 309–317.

Bright, A., Reynolds, G.R., Innes, J., Waas, J.R., 2003. Effects of motorised boat passes on the time budgets of New Zealand dabchick, *Poliocephalus rufopectus*. Wildl. Res. 30, 237–244.

Brivio, F., Grignolio, S., Brambilla, A., Apollonio, M., 2014. Intra-sexual variability in feeding behaviour of a mountain ungulate: size matters. Behav. Ecol. Sociobiol. 68, 1649–1660.

Brock, V.E., Riffenburgh, R.H., 1960. Fish schooling: a possible factor in reducing predation. J. Cons. Int. Explor. Mer. 25, 307–317.

Broder, E.D., Angeloni, L.M., 2014. Predator-induced phenotypic plasticity of laterality. Anim. Behav. 98, 125–130.

Brookes, M., 2003. Extreme Measures: The Dark Visions and Bright Ideas of Francis Galton. Bloomsbury, New York.

Broom, M., Ruxton, G.D., 1998. Evolutionarily stable stealing – game theory applied to kleptoparasitism. Behav. Ecol. 9, 397–403.

Brotons, L., Orell, M., Lahti, K., Koivula, K., 2000. Age-related microhabitat segregation in willow tit *Parus montanus* winter flocks. Ethology 106, 993–1005.

Brown, C.L., Hardy, A.R., Barber, J.R., Fristrup, K.M., Crooks, K.R., Angeloni, L.M., 2012. The effect of human activities and their associated noise on ungulate behavior. Plos One 7, e40505.

Brown, G.E., Godin, J.G.J., Pedersen, J., 1999a. Fin-flicking behaviour: a visual antipredator alarm signal in a characin fish. Anim. Behav. 58, 469–475.

Brown, J.S., 1988. Patch use as an indicator of habitat preference, predation risk, and competition. Behav. Ecol. Sociobiol. 22, 37–47.

Brown, J.S., Laundré, J.W., Gurung, M., 1999b. The ecology of fear: optimal foraging, game theory, and trophic interactions. J. Mammal. 80, 385–399.

Bshary, R., Noë, R., 1997. Red colobus and Diana monkeys provide mutual protection against predators. Anim. Behav. 54, 1461–1474.

Bugnyar, T., Kotrschal, K., 2002. Scrounging tactics in free-ranging ravens, *Corvus corax*. Ethology 108, 993–1009.

Bull, C.M., Pamula, Y., 1998. Enhanced vigilance in monogamous pairs of the lizard, *Tiliqua rugosa*. Behav. Ecol. 9, 452–455.

Burger, J., 1991. Foraging behavior and the effect of human disturbance on the piping plover (*Charadrius melodus*). J. Coast. Res. 7, 39–52.

Burger, J., 1994. The effect of human disturbance on foraging behavior and habitat use in piping plover (*Charadrius melodus*). Estuaries 17, 695–701.

Burger, J., 1997. Vigilance behavior and chickens: differences among status and location. Appl. Anim. Behav. Sci. 54, 345–350.

Burger, J., Gochfeld, M., 1988. Effects of group size and sex on vigilance in ostriches (*Struthio camelus*): antipredator strategy or mate competition? Ostrich 59, 14–20.

Burger, J., Gochfeld, M., 1991a. Human activity influence and diurnal and nocturnal foraging of sanderlings (*Calidris alba*). Condor 93, 259–265.

Burger, J., Gochfeld, M., 1991b. Vigilance and feeding behaviour in large feeding flocks of laughing gulls, *Larus atricilla*, on Delaware Bay. Estuar. Coast. Shelf Sci. 32, 207–212.

Burger, J., Gochfeld, M., 1994. Vigilance in African mammals: differences among mothers, other females, and males. Behaviour 131, 153–169.

Burger, J., Gochfeld, M., 1998. Effects of ecotourists on bird behaviour at Loxahatchee National Wildlife Refuge, Florida. Environ. Conserv. 25, 13–21.

Burger, J., Gochfeld, M., 2001. Effect of human presence on foraging behavior of Sandhill cranes (*Grus canadensis*) in Nebraska. Bird Behav. 2, 81–87.

Burger, J., Safina, C., Gochfeld, M., 2000. Factors affecting vigilance in springbok: importance of vegetative cover, location in herd, and herd size. Acta Ethol. 2, 97–104.

Butler, S.J., Whittingham, M.J., Quinn, J.L., Cresswell, W., 2005. Quantifying the interaction between food density and habitat structure in determining patch selection. Anim. Behav. 69, 337–343.

Byers, J.A., 1997. American Pronghorn: Social Adaptations and the Ghosts of Predation Past. University of Chicago Press, Chicago.

Byrkjedal, I., Kålås, J.A., 1983. Plover's page turn into plover's parasite: a look at the dunlin/plover association. Ornis Fennica 60, 10–15.

Cade, B.S., Noon, B.R., 2003. A gentle introduction to quantile regression for ecologists. Front. Ecol. Environ. 1, 412–420.

Cade, B.S., Terrell, J.W., Schroeder, R.L., 1999. Estimating effects of limiting factors with regression quantiles. Ecology 80, 311–323.

Caine, N.G., Marra, S.L., 1988. Vigilance and social organization in two species of primates. Anim. Behav. 36, 897–904.

Caithamer, D.F., Gates, R.J., Tacha, T.C., 1996. A comparison of diurnal time budgets from paired interior Canada geese with and without offspring. J. Field Ornithol. 67, 105–113.

Caldwell, G.S., 1986. Predation as a selective force on foraging herons: effects of plumage color and flocking. Auk 103, 494–505.

Cameron, E.S., 1908. Observations on the golden eagle in Montana. Auk 25, 251–268.

Cameron, E.Z., Du Toit, J.T., 2005. Social influences on vigilance behaviour in giraffes, *Giraffa camelopardalis*. Anim. Behav. 69, 1337–1344.

Caraco, T., 1979a. Time budgeting and group size: a test of theory. Ecology 60, 618–627.

Caraco, T., 1979b. Time budgeting and group size: a theory. Ecology 60, 611–617.

Caraco, T., 1982. Flock size and the organisation of behavioral sequences in juncos. Condor 84, 101–105.

Caraco, T., Martindale, S., Pulliam, H.R., 1980a. Avian flocking in the presence of a predator. Nature 285, 400–401.

Caraco, T., Martindale, S., Pulliam, H.R., 1980b. Avian time budgets and distance from cover. Auk 97, 872–875.

Carbone, C., Thompson, W.A., Zadorina, L., Rowcliffe, J.M., 2003. Competition, predation risk and patterns of flock expansion in barnacle geese (*Branta leucopsis*). J. Zool. 259, 301–308.

Carey, H.V., Moore, P., 1986. Foraging and predation risk in yellow-bellied marmots. Am. Midl. Nat. 116, 267–275.

Caro, T.M., 1986. The functions of stotting in Thomson's gazelles: some tests of the predictions. Anim. Behav. 34, 663–684.

Caro, T.M., 1987. Cheetah mothers' vigilance: looking out for prey or predators? Behav. Ecol. Sociobiol. 20, 351–361.

Caro, T.M., 2005. Antipredator Defenses in Birds and Mammals. University of Chicago Press, Chicago.

Caro, T.M., Graham, C.M., Stoner, C.J., Vargas, J.K., 2004. Adaptive significance of antipredator behaviour in artiodactyls. Anim. Behav. 67, 205–228.

Carr, J.M., Lima, S.L., 2010. High wind speeds decrease the responsiveness of birds to potentially threatening moving stimuli. Anim. Behav. 80, 215–220.

Carr, J.M., Lima, S.L., 2014. Wintering birds avoid warm sunshine: predation and the costs of foraging in sunlight. Oecologia 174, 713–721.

Carrascal, L.M., Diaz, J.A., Huertas, D.L., Mozetich, I., 2001. Behavioral thermoregulation by treecreepers: trade-off between saving energy and reduced crypsis. Ecology 82, 1642–1654.

Carrascal, L.M., Moreno, E., 1992. Proximal costs and benefits of heterospecific social foraging in the great tit, *Parus major*. Can. J. Zool. 70, 1947–1952.

Carro, M.E., Fernandez, G.J., 2009. Scanning pattern of greater rheas, *Rhea americana*: collective vigilance would increase the probability of detecting a predator. J. Ethol. 27, 429–436.

Carter, A.J., Goldizen, A.W., Tromp, S.A., 2010. Agamas exhibit behavioral syndromes: bolder males bask and feed more but may suffer higher predation. Behav. Ecol. 21, 655–661.

Carter, A.J., Pays, O., Goldizen, A.W., 2009. Individual variation in the relationship between vigilance and group size in eastern grey kangaroos. Behav. Ecol. Sociobiol. 64, 237–245.

Carter, K., Goldizen, A.W., 2003. Habitat choice and vigilance behaviour of brush-tailed rock-wallabies (*Petrogale penicillata*) within their nocturnal foraging ranges. Wildl. Res. 30, 355–364.

Cary, M., 1901. Birds of the Black Hills. Auk 18, 231–238.

Casas, F., Mougeot, F., Vinuela, J., Bretagnolle, V., 2009. Effects of hunting on the behaviour and spatial distribution of farmland birds: importance of hunting-free refuges in agricultural areas. Anim. Conserv. 12, 346–354.

Cassini, M.H., 1991. Foraging under predation risk in the wild guinea pig *Cavia aperea*. Oikos 62, 20–24.

Castellanos, I., Barbosa, P., 2006. Evaluation of predation risk by a caterpillar using substrate-borne vibrations. Anim. Behav. 72, 461–469.

Catterall, C.P., Elgar, M.A., Kikkawa, J., 1992. Vigilance does not covary with group size in an island population of silvereyes (*Zosterops lateralis*). Behav. Ecol. 3, 207–210.

Cézilly, F., Brun, B., 1989. Surveillance et picorage chez la tourterelle rieuse, *Streptopelia risoria*: effets de la présence d'un congénère et de la dispersion des graines. Behaviour 110, 146–160.

Cézilly, F., Keddar, I., 2012. Vigilance and food intake rate in paired and solitary Zenaida doves *Zenaida aurita*. Ibis 154, 161–166.

Chace, J.F., Walsh, J.J., 2006. Urban effects on native avifauna: a review. Landscape Urban Plan. 74, 46–69.

Chamberlain, S.A., Hovick, S.M., Dibble, C.J., Rasmussen, N.L., Van Allen, B.G., Maitner, B.S., Ahern, J.R., Bell-Dereske, L.P., Roy, C.L., Meza-Lopez, M., Carrillo, J., Siemann, E., Lajeunesse, M.J., Whitney, K.D., 2012. Does phylogeny matter? Assessing the impact of phylogenetic information in ecological meta-analysis. Ecol. Lett. 15, 627–636.

Chamove, A.S., Hosey, G.R., Schaetzel, P., 1988. Visitors excite primates in zoos. Zoo Biol. 7, 359–369.

Chan, A.A.Y.-H., Giraldo-Perez, P., Smith, S., Blumstein, D.T., 2010. Anthropogenic noise affects risk assessment and attention: the distracted prey hypothesis. Biol. Lett. 6, 458–461.

Chance, M.R.A., 1956. Social structure of a colony of *Macaca mulatta*. Br. J. Anim. Behav. 4, 1–13.

Chance, M.R.A., 1967. Attention structure as the basis of primate rank orders. Man 2, 503–518.

Chang, S.W.C., Platt, M.L., 2014. Oxytocin and social cognition in rhesus macaques: implications for understanding and treating human psychopathology. Brain Res. 1580, 57–68.

Chapman, C.A., Chapman, L.J., 1996. Mixed-species primate groups in the Kibale forest: ecological constraints on association. Int. J. Primatol. 17, 31–50.

Charmandari, E., Tsigos, C., Chrousos, G., 2004. Endocrinology of the stress response. Annu. Rev. Physiol. 67, 259–284.

Childress, M.J., Lung, M.A., 2003. Predation risk, gender and the group size effect: does elk vigilance depend upon the behaviour of conspecifics? Anim. Behav. 66, 389–398.

Choy, K.H.C., Yu, J., Hawkes, D., Mayorov, D.N., 2012. Analysis of vigilant scanning behavior in mice using two-point digital video tracking. Psychopharmacology 221, 649–657.

Christman, M.C., Leone, E.H., 2007. Statistical aspects of the analysis of group size effects in confined animals. Appl. Anim. Behav. Sci. 103, 265–283.

Cimprich, D.A., Grubb, T.C., 1994. Consequences for Carolina chickadees of foraging with tufted titmice in winter. Ecology 75, 1615–1625.

Ciuti, S., Northrup, J.M., Muhly, T.B., Simi, S., Musiani, M., Pitt, J.A., Boyce, M.S., 2012. Effects of humans on behaviour of wildlife exceed those of natural predators in a landscape of fear. Plos One 7, e50611.

Clair, C.C.S., Forrest, A., 2009. Impacts of vehicle traffic on the distribution and behaviour of rutting elk, *Cervus elaphus*. Behaviour 146, 393–413.

Clark, C.W., Mangel, M., 1986. The evolutionary advantages of group foraging. Theor. Popul. Biol. 30, 45–75.

Clarke, J.L., Jones, M.E., Jarman, P.J., 1989. A day in the life of a kangaroo: activities and movements of eastern grey kangaroos *Macropus giganteus* at Wallaby Creek. In: Grigg, G.C., Jarman, P.J., Hume, I. (Eds.), Kangaroos, Wallabies and Rat-Kangaroos. Surrey Beatty and Sons, New South Wales.

Claxton, A.M., 2011. The potential of the human–animal relationship as an environmental enrichment for the welfare of zoo-housed animals. Appl. Anim. Behav. Sci. 133, 1–10.

Clutton-Brock, T., Sheldon, B.C., 2010. Individuals and populations: the role of long-term, individual-based studies of animals in ecology and evolutionary biology. Trends Ecol. Evol. 25, 562–573.

Clutton-Brock, T.H., O'Riain, M.J., Brotherton, P.N.M., Gaynor, D., Kansky, R., Griffin, A.S., Manser, M., 1999. Selfish sentinels in cooperative mammals. Science 284, 1640–1644.

Colagross, A.M.L., Cockburn, A., 1993. Vigilance and grouping in the eastern grey kangaroo, *Macropus giganteus*. Aust. J. Zool. 41, 325–334.

Coleman, S.W., 2008. Mourning dove (*Zenaida macroura*) wing-whistles may contain threat-related information for con- and hetero-specifics. Naturwissenschaften 95, 981–986.

Conomy, J.T., Collazo, J.A., Dubovsky, J.A., Fleming, W.J., 1998. Dabbling duck behavior and aircraft activity in coastal North Carolina. J. Wildl. Manage. 62, 1127–1134.

Coolen, I., Giraldeau, L.-A., 2003. Incompatibility between antipredatory vigilance and scrounger tactic in nutmeg manikins, *Lonchura punctulata*. Anim. Behav. 66, 657–664.

Coolen, I., Giraldeau, L.-A., Lavoie, M., 2001. Head position as an indicator of producer and scrounger tactics in a ground-feeding bird. Anim. Behav. 61, 895–903.

Cords, M., 1990. Vigilance and mixed-species association of some East African forest monkeys. Behav. Ecol. Sociobiol. 26, 297–300.

Coss, R.G., 1999. Effects of relaxed natural selection on the evolution of behavior. In: Foster, S.A., Endler, J.A. (Eds.), Geographic Variation in Behavior: Perspectives on Evolutionary Mechanisms. Oxford University Press, Oxford, pp. 180–208.

Couchoux, C., Cresswell, W., 2012. Personality constraints versus flexible antipredation behaviors: how important is boldness in risk management of redshanks (*Tringa totanus*) foraging in a natural system? Behav. Ecol. 23, 290–301.

Cowlishaw, G., 1994. Vulnerability to predation in baboon populations. Behaviour 131, 293–304.

Cowlishaw, G., Lawes, M.J., Lightbody, M., Martin, A., Pettifor, R., Rowcliffe, J.M., 2004. A simple rule for the costs of vigilance: empirical evidence from a social forager. Proc. R. Soc. Lond. B 271, 27–33.

Creel, S., Schuette, P., Christianson, D., 2014. Effects of predation risk on group size, vigilance, and foraging behavior in an African ungulate community. Behav. Ecol. 25, 773–784.

Creel, S., Winnie, J.A., 2005. Responses of elk herd size to fine-scale spatial and temporal variation in the risk of predation by wolves. Anim. Behav. 69, 1181–1189.

Creel, S., Winnie Jr., J.A., Christianson, D., Liley, S., 2008. Time and space in general models of antipredator response: tests with wolves and elk. Anim. Behav. 76, 1139–1146.

Cresswell, W., 1994. Flocking is an effective anti-predation strategy in redshanks, *Tringa totanus*. Anim. Behav. 47, 433–442.

Cresswell, W., Hilton, G.M., Ruxton, G.D., 2000. Evidence for a rule governing the avoidance of superfluous escape flights. Proc. R. Soc. Lond. B 267, 733–737.

Cresswell, W., Quinn, J.L., Whittingham, M.J., Butler, S., 2003. Good foragers can also be good at detecting predators. Proc. R. Soc. Lond. B 270, 1069–1076.

Cripps, J.K., Wilson, M.E., Elgar, M.A., Coulson, G., 2011. Experimental manipulation of fertility reveals potential lactation costs in a free-ranging marsupial. Biol. Lett. 7, 859–862.

Crook, J.H., 1965. The adaptive significance of avian social organisations. Symp. Zool. Soc. Lond. 14, 181–218.

Crook, J.H., 1970. Social Behaviour in Birds and Mammals. Academic Press, New York.

Crooks, K.R., 2002. Relative sensitivities of mammalian carnivores to habitat fragmentation. Conserv. Biol. 16, 488–502.

Crosmary, W.G., Makumbe, P., Côté, S.D., Fritz, H., 2012. Vulnerability to predation and water constraints limit behavioural adjustments of ungulates in response to hunting risk. Anim. Behav. 83, 1367–1376.

Da Silva, J., Terhune, J.M., 1988. Harbour seal grouping as an anti-predator strategy. Anim. Behav. 36, 1309–1316.

Dadda, M., Bisazza, A., 2006. Does brain asymmetry allow efficient performance of simultaneous tasks? Anim. Behav. 72, 523–529.

Dalmau, A., Ferret, A., Manteca, X., 2010. Vigilance behavior of Pyrenean chamois *Rupicapra pyrenaica pyrenaica*: effect of sex and position in the herd. Curr. Zool. 56, 232–237.

Danchin, E., Giraldeau, L.A., Valone, T.J., Wagner, R.H., 2004. Public information: from nosy neighbors to cultural evolution. Science 305, 487–491.

Dannock, R.J., Blomberg, S.P., Goldizen, A.W., 2013. Individual variation in vigilance in female eastern grey kangaroos. Aust. J. Zool. 61, 312–319.

Darling, F.F., 1937. A Herd of Red Deer. Oxford University Press, Oxford.

Darwin, C., 1872. The Expression of the Emotions in Man and Animals. John Murray, London.

Datta, S.B., Beauchamp, G., 1991. Demography and female dominance patterns in primates 1 mother–daughter and sister–sister relationships. Am. Nat. 138, 201–226.

Davey, G., 2007. Visitors' effects on the welfare of animals in the zoo: a review. J. Appl. Anim. Welf. Sci. 10, 169–183.

Davidson, G.L., Butler, S., Fernández-Juricic, E., Thornton, A., Clayton, N.S., 2014. Gaze sensitivity: function and mechanisms from sensory and cognitive perspectives. Anim. Behav. 87, 3–15.

Davies, D.R., Parasuraman, R., 1982. The Psychology of Vigilance. Academic Press, New York.

Dawkins, M.S., 2002. What are birds looking at? Head movements and eye use in chickens. Anim. Behav. 63, 991–998.

de Ruiter, J.R., 1986. The influence of group size on predator scanning and foraging behaviour of wedge-capped capuchin monkeys (Cebus olivaceus). Behaviour 98, 240–258.

Dehn, M.M., 1990. Vigilance for predators: detection and dilution effects. Behav. Ecol. Sociobiol. 26, 337–342.

Dekker, D., Dekker, I., Christie, D., Ydenberg, R., 2011. Do staging semipalmated sandpipers spend the high-tide period in flight over the ocean to avoid falcon attacks along shore? Waterbirds 34, 195–201.

Desportes, J.-P., Cézilly, F., Gallo, A., 1991a. Modelling and analysing vigilance behaviour. Acta Oecol. 12, 227–236.

Desportes, J.-P., Gallo, A., Cézilly, F., 1991b. Effet de la familiarisation avec l'environnement sur le comportement de vigilance de la tourterelle rieuse Streptopelia risoria. Behav. Proc. 24, 177–183.

Desportes, J.-P., Metcalfe, N.B., Brun, B., Cézilly, F., 1990a. Spectral analysis of series of scan durations and their relationship with inter-scan intervals. Ethology 85, 43–50.

Desportes, J.P., Brun, B., Cézilly, F., 1990b. Vigilant behaviour: predictability or randomness? Spectral analysis of series of scan durations and their relationship with inter-scan intervals. Ethology 85, 43–50.

Desportes, J.P., Metcalfe, N.B., Cézilly, F., Lauvergnon, G., Kervella, C., 1989. Tests of sequential randomness of vigilant behaviour using spectral analysis. Anim. Behav. 38, 771–777.

Desportes, J.P., Metcalfe, N.B., Popp, J.W., Meyer, R.M., Gallo, A., Cézilly, F., 1990c. The predictability and patterns of vigilant behaviour. Behav. Proc. 22, 41–46.

Devereux, C.L., Whittingham, M.J., Fernández-Juricic, E., Vickery, J.A., Krebs, J.R., 2006. Predator detection and avoidance by starlings under differing scenarios of predation risk. Behav. Ecol. 17, 303–309.

Di Blanco, Y., Hirsch, B.T., 2006. Determinants of vigilance behavior in the ring-tailed coati (Nasua nasua): the importance of within-group spatial position. Behav. Ecol. Sociobiol. 61, 173–182.

Diamond, J.M., 1981. Mixed-species foraging groups. Nature 292, 408–409.

Dias, R.I., 2006. Effects of position and flock size on vigilance and foraging behaviour of the scaled dove Columbina squammata. Behav. Proc. 73, 248–252.

Diaz, J.A., Asensio, B., 1991. Effects of group size and distance to protective cover on the vigilance behaviour of black-billed magpies (Pica pica). Bird Study 38, 38–41.

Dijk, D.-J., Lockley, S.W., 2002. Integration of human sleep-wake regulation and circadian rhythmicity. J. Appl. Physiol. 92, 852–862.

Dimond, S., Lazarus, J., 1974. The problem of vigilance in animal life. Brain Behav. Evol. 9, 60–79.

Djagoun, C.A.M.S., Djossa, B.A., Mensah, G.A., Sinsin, B.A., 2013. Vigilance efficiency and behaviour of Bohor reedbuck Redunca redunca (Pallas 1767) in a savanna environment of Pendjari Biosphere Reserve (Northern Benin). Mamm. Study 38, 81–89.

Domènech, J., Senar, J.C., 1999. Are foraging serin Serinus serinus females more vigilant than males? The effect of sex ratio. Ardea 87, 277–284.

Dominguez, J., 2002. Biotic and abiotic factors affecting the feeding behavior of the black-tailed godwit. Waterbirds 25, 393–400.

Dominguez, J., 2003. Sleeping and vigilance in black-tailed godwit. J. Ethol. 21, 57–60.

Dominguez, J., Vidal, M., 2007. Vigilance behaviour of preening black-tailed godwit *Limosa limosa* in roosting flocks. Ardeola 54, 227–235.

Dreiss, A.N., Calcagno, M., van den Brink, V., Laurent, A., Almasi, B., Jenni, L., Roulin, A., 2013. The vigilance components of begging and sibling competition. J. Avian Biol. 44, 359–368.

Duchesne, M., Côté, S.D., Barrette, C., 2000. Responses of woodland caribou to winter ecotourism in the Charlevoix Biosphere Reserve. Can. Biol. Conserv. 96, 311–317.

Dukas, R., Kamil, A.C., 2000. The cost of limited attention in blue jays. Behav. Ecol. 11, 502–506.

Dunbar, R.I.M., Cornah, L., Daly, F.J., Bowyer, K.M., 2002. Vigilance in human groups: a test of alternative hypotheses. Behaviour 139, 695–711.

Duncan, R.D., Jenkins, S.H., 1998. Use of visual cues in foraging by a diurnal herbivore Belding's ground squirrel. Can. J. Zool. 76, 1766–1770.

Dunham, J.B., Cade, B.S., Terrell, J.W., 2002. Influences of spatial and temporal variation of fish-habitat relationships defined by regression quantiles. Trans. Am. Fish. Soc. 131, 86–98.

Dyck, M.G., Baydack, R.K., 2004. Vigilance behaviour of polar bears (*Ursus maritimus*) in the context of wildlife-viewing activities at Churchill, Manitoba, Canada. Biol. Conserv. 116, 343–350.

Ebensperger, L.A., Hurtado, M.A.J., Ramos-Jiliberto, R., 2006. Vigilance and collective detection of predators in degus (*Octodon degus*). Ethology 112, 879–887.

Ebensperger, L.A., Hurtado, M.J., 2005a. On the relationship between herbaceous cover and vigilance activity of degus (*Octodon degus*). Ethology 111, 593–608.

Ebensperger, L.A., Hurtado, M.J., 2005b. Seasonal changes in the time budget of degus (*Octodon degus*). Behaviour 142, 91–112.

Ebitz, R.B., Pearson, J.M., Platt, M.L., 2014. Pupil size and social vigilance in rhesus macaques. Front. Neurosci. 8, e100.

Ebitz, R.B., Watson, K.K., Platt, M.L., 2013. Oxytocin blunts social vigilance in the rhesus macaque. Proc. Natl. Acad. Sci. 110, 11630–11635.

Edwards, A.M., Best, E.C., Blomberg, S.P., Goldizen, A.W., 2013. Individual traits influence vigilance in wild female eastern grey kangaroos. Aust. J. Zool. 61, 332–341.

Eilam, D., Izhar, R., Mort, J., 2011. Threat detection: behavioral practices in animals and humans. Neurosci. Biobehav. Rev. 35, 999–1006.

Ekman, J., 1987. Exposure and time use in willow tit flocks: the cost of subordination. Anim. Behav. 35, 508–514.

Ekman, J., Askenmo, C.E.H., 1984. Social rank and habitat use in willow tit groups. Anim. Behav. 32, 508–514.

Ekman, J., Cederholm, G., Askemo, C., 1981. Spacing and survival in winter groups of willow tit (*Parus montanus*) and crested tit (*P. cristatus*): a removal study. Ornis Scand. 10, 55–68.

Elcavage, P., Caraco, T., 1983. Vigilance behaviour in house sparrow flocks. Anim. Behav. 31, 303–304.

Eldar, E., Cohen, J.D., Niv, Y., 2013. The effects of neural gain on attention and learning. Nat. Neurosci. 16, 1146–1153.

Elgar, M.A., 1986a. The establishment of foraging flocks in house sparrows: risk of predation and daily temperature. Behav. Ecol. Sociobiol. 19, 433–438.

Elgar, M.A., 1986b. Scanning, pecking and alarm flights in house sparrows. Anim. Behav. 34, 1892–1894.

Elgar, M.A., 1989. Predator vigilance and group size in mammals and birds. Biol. Rev. 64, 13–33.

Elgar, M.A., Burren, P.J., Posen, M., 1984. Vigilance and perception of flock size in foraging house sparrows *Passer domesticus* L. Behaviour 90, 215–223.

Elgar, M.A., Catterall, C.P., 1982. Flock size and feeding efficiency in house sparrows. Emu 82, 109–111.

Elliott, D.G., 1913. A review of the primates. American Museum of Natural History.

Embar, K., Kotler, B.P., Mukherjee, S., 2011. Risk management in optimal foragers: the effect of sightlines and predator type on patch use, time allocation, and vigilance in gerbils. Oikos 120, 1657–1666.

Emery, N.J., 2000. The eyes have it: the neuroethology, function and evolution of social gaze. Neurosci. Biobehav. Rev. 24, 581–604.

Endler, J.A., 1991. Interactions between predators and prey. In: Krebs, J.R., Davies, N.B. (Eds.), Behavioural Ecology. Blackwell Scientific, Oxford, pp. 169–196.

Engelhard, G.H., Baarspul, A.N.J., Broekman, M., Creuwels, J.C.S., Reijnders, P.J.H., 2002. Human disturbance, nursing behaviour, and lactational pup growth in a declining southern elephant seal (*Mirounga leonina*) population. Can. J. Zool. 80, 1876–1886.

Evers, E., de Vries, H., Spruijt, B.M., Sterck, E.H.M., 2012. Look before you leap – individual variation in social vigilance shapes socio-spatial group properties in an agent-based model. Behav. Ecol. Sociobiol. 66, 931–945.

Fairbanks, B., Dobson, F.S., 2007. Mechanisms of the group-size effect on vigilance in Columbian ground squirrels: dilution versus detection. Anim. Behav. 73, 115–123.

Fanson, K.V., Fanson, B.G., Brown, J.S., 2011. Using path analysis to explore vigilance behavior in the rock hyrax (*Procavia capensis*). J. Mammal. 92, 78–85.

Favreau, F.-R., Goldizen, A.W., Fritz, H., Blomberg, S.P., Best, E.C., Pays, O., 2014. Within-population differences in personality and plasticity in the trade-off between vigilance and foraging in kangaroos. Anim. Behav. 92, 175–184.

Favreau, F.-R., Goldizen, A.W., Pays, O., 2010. Interactions among social monitoring, anti-predator vigilance and group size in eastern grey kangaroos. Proc. R. Soc. Lond. B 277, 2089–2095.

Favreau, F.R., Jarman, P.J., Goldizen, A.W., Dubot, A.L., Source, S., Pays, O., 2009. Vigilance in a solitary marsupial, the common wombat (*Vombatus ursinus*). Aust. J. Zool. 57, 363–371.

Favreau, F.R., Pays, O., Goldizen, A.W., Fritz, H., 2013. Short-term behavioural responses of impalas in simulated antipredator and social contexts. Plos One 8, e84970.

Feare, C., Craig, A., 1998. Starlings and Mynas. Christopher Helm, London.

Ferguson, J.W.H., 1987. Vigilance behaviour in white-browed sparrow-weavers *Plocepasser mahali*. Ethology 76, 223–235.

Fernández-Juricic, E., 2012. Sensory basis of vigilance behavior in birds: synthesis and future prospects. Behav. Proc. 89, 143–152.

Fernández-Juricic, E., Beauchamp, G., 2008. An experimental analysis of spatial position effects on foraging and vigilance in brown-headed cowbird flocks. Ethology 114, 105–114.

Fernández-Juricic, E., Beauchamp, G., Bastain, B., 2007a. Group-size and distance-to-neighbour effects on feeding and vigilance in brown-headed cowbirds. Anim. Behav. 73, 771–778.

Fernández-Juricic, E., Beauchamp, G., Treminio, R., Hoover, M., 2011a. Making heads turn: association between head movements during vigilance and perceived predation risk in brown-headed cowbird flocks. Anim. Behav. 82, 573–577.

Fernández-Juricic, E., Deisher, M., Stark, A.C., Randolet, J., 2012. Predator detection is limited in microhabitats with high light intensity: an experiment with brown-headed cowbirds. Ethology 118, 341–350.

Fernández-Juricic, E., Delgado, J.A., Remacha, C., Jiménez, M.D., Garcia, V., Hori, K., 2009. Can a solitary avian species use collective detection? An assay in semi-natural conditions. Behav. Proc. 82, 67–74.

Fernández-Juricic, E., Erichsen, J.T., Kacelnik, A., 2004a. Visual perception and social foraging in birds. Trends Ecol. Evol. 19, 25–31.

Fernández-Juricic, E., Gall, M.D., Dolan, T., O'Rourke, C., Thomas, S., Lynch, J.R., 2011b. Visual systems and vigilance behaviour of two ground-foraging avian prey species: white-crowned sparrows and California towhees. Anim. Behav. 81, 705–713.

Fernández-Juricic, E., Gall, M.D., Dolan, T., Tisdale, V., Martin, G.R., 2008. The visual fields of two ground-foraging birds, House Finches and House Sparrows, allow for simultaneous foraging and anti-predator vigilance. Ibis 150, 779–787.

Fernández-Juricic, E., Kacelnik, A., 2004. Information transfer and gain in flocks: the effects of quality and quantity of social information at different neighbour distances. Behav. Ecol. Sociobiol. 55, 502–511.

Fernández-Juricic, E., Kerr, B., Bednekoff, P.A., Stephens, D.W., 2004b. When are two heads better than one? Visual perception and information transfer affect vigilance coordination in foraging groups. Behav. Ecol. 15, 898–906.

Fernández-Juricic, E., Kowalski, V., 2011. Where does a flock end from an information perspective? A comparative experiment with live and robotic birds. Behav. Ecol. 22, 1304–1311.

Fernández-Juricic, E., O'Rourke, C., Pitlik, T., 2010. Visual coverage and scanning behavior in two corvid species: American crow and Western scrub jay. J. Comp. Physiol. A Neuroethol. Sens. Neural Behav. Physiol. 196, 879–888.

Fernández-Juricic, E., Schroeder, N., 2003. Do variations in scanning behavior affect tolerance to human disturbance? Appl. Anim. Behav. Sci. 84, 219–234.

Fernández-Juricic, E., Siller, S., Kacelnik, A., 2004c. Flock density, social foraging, and scanning: an experiment with starlings. Behav. Ecol. 15, 371–379.

Fernández-Juricic, E., Smith, R., Kacelnik, A., 2005. Increasing the costs of conspecific scanning in socially foraging starlings affects vigilance and foraging behaviour. Anim. Behav. 69, 73–81.

Fernández-Juricic, E., Telleria, J.L., 2000. Effects of human disturbance on spatial and temporal feeding patterns of blackbird *Turdus merula* in urban parks in Madrid, Spain. Bird Study 47, 13–21.

Fernández-Juricic, E., Tran, E., 2007. Changes in vigilance and foraging behaviour with light intensity and their effects on food intake and predator detection in house finches. Anim. Behav. 74, 1381–1390.

Fernández-Juricic, E., Zollner, P.A., LeBlang, C., Westphal, L.M., 2007b. Responses of nestling black-crowned night herons (*Nycticorax nycticorax*) to aquatic and terrestrial recreational activities: a manipulative study. Waterbirds 30, 554–565.

Ferrari, M.C.O., Wisenden, B.D., Chivers, D.P., 2010. Chemical ecology of predator–prey interactions in aquatic ecosystems: a review and prospectus. Can. J. Zool. 88, 698–724.

Ferrière, R., Cazelles, B., Cézilly, F., Desportes, J.-P., 1996. Predictability and chaos in bird vigilant behaviour. Anim. Behav. 52, 457–472.

FitzGibbon, C.D., 1989. A cost to individual with reduced vigilance in groups of Thomson's gazelles hunted by cheetahs. Anim. Behav. 37, 508–510.

FitzGibbon, C.D., 1990. Mixed species grouping in Thompson's and Grant's gazelles: the antipredator benefits. Anim. Behav. 39, 1116–1126.

FitzGibbon, C.D., 1994. The costs and benefits of predator inspection behaviour in Thomson's gazelles. Behav. Ecol. Sociobiol. 34, 139–148.

FitzGibbon, C.D., Fanshawe, J., 1988. Stotting in Thomson's gazelles: an honest signal of condition. Behav. Ecol. Sociobiol. 23, 69–74.

Fitzpatrick, S., Bouchez, B., 1998. Effects of recreational disturbance on the foraging behaviour of waders on a rocky beach. Bird Study 45, 157–171.

Flemming, S.P., Chiasson, R.D., Smith, P.C., Austin-Smith, P.J., Bancroft, R.P., 1988. Piping plover status in Nova Scotia related to its reproductive and behavioral responses to human disturbance. J. Field Ornithol. 59, 321–330.

Flörcke, C., Engle, T.E., Grandin, T., Deesing, M.J., 2012. Individual differences in calf defence patterns in Red Angus beef cows. Appl. Anim. Behav. Sci. 139, 203–208.

Flörcke, C., Grandin, T., 2013. Loss of anti-predator behaviors in cattle and the increased predation losses by wolves in the Northern Rocky Mountains. Open J. Anim. Sci. 3, 248–253.

Fordyce, J.A., Agrawal, A.A., 2001. The role of plant trichomes and catterpillar group size on growth and defence of the pipevine swallowtail *Battus philenor*. J. Anim. Ecol. 70, 997–1005.

Forslund, P., 1993. Vigilance in relation to brood size and predator abundance in the barnacle goose, *Branta leucopsis*. Anim. Behav. 45, 965–973.

Fortin, D., Boyce, M.S., Merrill, E.H., 2004a. Multi-tasking by mammalian herbivores: overlapping processes during foraging. Ecology 85, 2312–2322.

Fortin, D., Boyce, M.S., Merrill, E.H., Fryxell, J.M., 2004b. Foraging costs of vigilance in large mammalian herbivores. Oikos 107, 172–180.

Foster-McDonald, N.S., Hygnstrom, S.E., Korte, S.P., 2006. Effects of a visual barrier fence on behaviour and movements of black-tailed prairie dogs. Wildl. Soc. B 34, 1169–1174.

Foster, W.A., Treherne, J.E., 1981. Evidence for the dilution effect in the selfish herd from fish predation on a marine insect. Nature 293, 466–467.

Fox, R.J., Donelson, J.M., 2014. Rabbitfish sentinels: first report of coordinated vigilance in conspecific marine fishes. Coral Reefs 33, 253.

Fragaszy, D.M., 1990. Sex and age differences in the organization of behavior in wedge-capped capuchins, *Cebus olivaceus*. Behav. Ecol. 1, 81–94.

Franklin, W.E., Lima, S.L., 2001. Laterality in avian vigilance: do sparrows have a favourite eye? Anim. Behav. 62, 879–885.

Freeberg, T.M., Krama, T., Vrublevska, J., Krams, I., Kullberg, C., 2014. Tufted titmouse (*Baeolophus bicolor*) calling and risk-sensitive foraging in the face of threat. Anim. Cogn. 17, 1341–1352.

Frid, A., 1997. Vigilance by female Dall's sheep: interaction between predation risk factors. Anim. Behav. 53, 799–808.

Frid, A., 2003. Dall's sheep responses to overflights by helicopter and fixed-wing aircraft. Biol. Conserv. 110, 387–399.

Frid, A., Dill, L.M., 2002. Human-caused disturbance stimuli as a form of predation risk. Conserv. Ecol. 6, 11.

Fritz, H., Guillemain, M., Durant, D., 2002. The cost of vigilance for intake rate in the mallard (*Anas platyrhynchos*): an approach through foraging experiments. Ethol. Ecol. Evol. 14, 91–97.

Fullard, J.H., 2001. Auditory sensitivity of Hawaiian moths (Lepidoptera : Noctuidae) and selective predation by the Hawaiian hoary bat (Chiroptera : *Lasiurus cinereus semotus*). Proc. R. Soc. Lond. B 268, 1375–1380.

Fuller, R.A., Bearhop, S., Metcalfe, N.B., Piersma, T., 2013. The effect of group size on vigilance in Ruddy Turnstones *Arenaria interpres* varies with foraging habitat. Ibis 155, 246–257.

Fusani, L., Beani, L., Lupo, C., DessiFulgheri, F., 1997. Sexually selected vigilance behaviour of the grey partridge is affected by plasma androgen levels. Anim. Behav. 54, 1013–1018.

Galicia, E., Baldassarre, G.A., 1997. Effects of motorized tourboats on the behavior of nonbreeding American flamingos in Yucatan. Mexico. Conserv. Biol. 11, 1159–1165.

Gallup, A.C., Chong, A., Kacelnik, A., Krebs, J.R., Couzin, I.D., 2014. The influence of emotional facial expressions on gaze-following in grouped and solitary pedestrians. Sci. Rep. 4, e5794.

Gallup, A.C., Hale, J.J., Sumpter, D.J.T., Garnier, S., Kacelnik, A., Krebs, J.R., Couzin, I.D., 2012. Visual attention and the acquisition of information in human crowds. Proc. Natl. Acad. Sci. 109, 7245–7250.

Galton, F., 1871. Gregariousness in cattle and men. MacMillan Mag. 23, 353–357.

Galton, F., 1883. Inquiries into Human Faculty and its Develoment. MacMillan, London.

Gaston, A.J., 1977. Social behaviour within groups of jungle babblers (*Turdoides striatus*). Anim. Behav. 25, 828–848.

Gauthier-Clerc, M., Tamisier, A., Cézilly, F., 1998. Sleep-vigilance trade-off in green-winged teals (*Anas crecca crecca*). Can. J. Zool. 76, 2214–2218.

Gauthier-Clerc, M., Tamisier, A., Cézilly, F., 2002. Vigilance while sleeping in the breeding pochard *Aythya ferina* according to sex and age. Bird Study 49, 300–303.

Gauthier, G., Tardif, J., 1991. Female feeding and male vigilance during nesting in greater snow geese. Condor 93, 701–711.

Gavin, S.D., Komers, P.E., 2006. Do pronghorn (*Antilocapra americana*) perceive roads as a predation risk? Can. J. Zool. 84, 1775–1780.

Gaynor, K.M., Cords, M., 2012. Antipredator and social monitoring functions of vigilance behaviour in blue monkeys. Anim. Behav. 84, 531–537.

Ge, C., Beauchamp, G., Li, Z., 2011. Coordination and synchronisation of anti-predation vigilance in two crane species. Plos One 6, e26447.

Gende, S.M., Quinn, T.P., 2004. The relative importance of prey density and social dominance in determining energy intake by bears feeding on Pacific salmon. Can. J. Zool. 82, 75–85.

Gerkema, M.P., Verhulst, S., 1990. Warning against an unseen predator: a functional aspect of synchronous feeding in the common vole, *Microtus arvalis*. Anim. Behav. 40, 1169–1176.

Ghosal, R., Venkataraman, A., 2013. An adaptive system of vigilance in spotted deer (*Axis axis*) herds in response to predation. Curr. Sci. 104, 768–771.

Giese, M., Riddle, M., 1999. Disturbance of emperor penguin *Aptenodytes forsteri* chicks by helicopters. Polar Biol. 22, 366–371.

Gill, J.A., 2007. Approaches to measuring the effects of human disturbance on birds. Ibis 149, 9–14.

Gill, J.A., Sutherland, W.J., Watkinson, A.R., 1996. A method to quantify the effects of human disturbance on animal populations. J. Appl. Ecol. 33, 786–792.

Giraldeau, L.-A., Caraco, T., 2000. Social Foraging Theory. Princeton University Press, Princeton.

Girard-Buttoz, C., Heistermann, M., Rahmi, E., Agil, M., Fauzan, P.A., Engelhardt, A., 2014. Costs of and investment in mate-guarding in wild long-tailed macaques (*Macaca fascicularis*): influences of female characteristics and male-female social bonds. Int. J. Primatol. 35, 701–724.

Godin, J.-G.J., Classon, L.J., Abrahams, M.V., 1988. Group vigilance and shoal size in a small characin fish. Behaviour 104, 29–40.

Goldsmith, A.E., 1990. Vigilance behavior of pronghorns in different habitats. J. Mammal. 71, 460–462.

Goodale, E., Kotagama, S.W., 2005. Alarm calling in Sri Lankan mixed-species bird flocks. Auk 122, 108–120.

Goss-Custard, J.D., 1970. Feeding dispersion in some over-wintering wading birds. In: Crook, J.H. (Ed.), Social Behaviour in Birds and Mammals. Academic Press, New York, pp. 3–35.

Goss-Custard, J.D., Cayford, J.T., Lea, S.G., 1999. Vigilance during food handling by oystercatchers *Haematopus ostralegus* reduces the chances of losing prey to kleptoparasites. Ibis 141, 368–376.

Gosselin-Ildari, A.D., Koenig, A., 2012. The effects of group size and reproductive status on vigilance in captive *Callithrix jacchus*. Am. J. Primatol. 74, 613–621.

Gothard, K.M., Battaglia, F.P., Erickson, C.A., Spitler, K.M., Amaral, D.G., 2007. Neural responses to facial expression and face identity in the monkey amygdala. J. Neurophysiol. 97, 1671–1683.

Götmark, F., 1993. Conspicuous coloration in male birds is favoured by predation in some species and disfavoured in others. Proc. R. Soc. Lond. B 253, 143–146.

Gould, L., 1996. Vigilance behavior during the birth and lactation season in naturally occurring ring-tailed lemurs (*Lemur catta*) at the Beza-Mahafaly Reserve, Madagascar. Int. J. Primatol. 17, 331–347.

Gould, L., Fedigan, L.M., Rose, L.M., 1997. Why be vigilant? The case of the alpha animal. Int. J. Primatol. 18, 401–414.

Grand, T.C., Dill, L.M., 1999. The effect of group size on the foraging behaviour of juvenile coho salmon: reduction of predation risk or increased competition? Anim. Behav. 58, 443–451.

Granquist, S.M., Sigurjonsdottir, H., 2014. The effect of land based seal watching tourism on the haul-out behaviour of harbour seals (*Phoca vitulina*) in Iceland. Appl. Anim. Behav. Sci. 156, 85–93.

Green, A.J., Fox, A.D., Hughes, B., Hilton, G.M., 1999. Time-activity budgets and site selection of white-headed ducks *Oxyura leucocephala* at Burdur Lake, Turkey in late winter. Bird Study 46, 62–73.

Greig-Smith, P.W., 1981. The role of alarm responses in the formation of mixed-species flocks of heathland birds. Behav. Ecol. Sociobiol. 8, 7–10.

Griesser, M., 2003. Nepotistic vigilance behavior in Siberian jay parents. Behav. Ecol. 14, 246–250.

Griesser, M., Nystrand, M., 2009. Vigilance and predation of a forest-living bird species depend on large-scale habitat structure. Behav. Ecol. 20, 709–715.

Griffin, A.S., Blumstein, D.T., Evans, C.S., 2000. Training captive-bred or translocated animals to avoid predators. Conserv. Biol. 14, 1317–1326.

Griffin, A.S., Evans, C.S., Blumstein, D.T., 2001. Learning specificity in acquired predator recognition. Anim. Behav. 62, 577–589.

Griffin, S.C., Valois, T., Taper, M.L., Mills, L.S., 2007. Effects of tourists on behavior and demography of Olympic marmots. Conserv. Biol. 21, 1070–1081.

Grinnell, J., 1903. Call notes of the bushtit. Condor 5, 85–87.

Guillemain, M., Caldow, R.W.G., Hodder, K.H., Goss-Custard, J.D., 2003. Increased vigilance of paired males in sexually dimorphic species: distinguishing between alternative explanations in wintering Eurasian wigeon. Behav. Ecol. 14, 724–729.

Guillemain, M., Duncan, P., Fritz, H., 2001. Switching to a feeding method that obstructs vision increases head-up vigilance in dabbling ducks. J. Avian Biol. 32, 345–350.

Gustafsson, L., 1988. Foraging behaviour of individual coal tits, *Parus ater*, in relation to their age, sex and morphology. Anim. Behav. 36, 696–704.

Hall, C.L., Fedigan, L.M., 1997. Spatial benefits afforded by high rank in white-faced capuchins. Anim. Behav. 53, 1069–1082.

Hall, K.R.L., 1960. Social vigilance behaviour of the Chacma baboon, *Papio ursinus*. Behaviour 16, 261–294.

Hall, L.K., Day, C.C., Westover, M.D., Edgel, R.J., Larsen, R.T., Knight, R.N., McMillan, B.R., 2013. Vigilance of kit foxes at water sources: a test of competing hypotheses for a solitary carnivore subject to predation. Behav. Proc. 94, 76–82.

Halofsky, J.S., Ripple, W.J., 2008. Fine-scale predation risk on elk after wolf reintroduction in Yellowstone National Park, USA. Oecologia 155, 869–877.

Hamel, S., Côté, S.D., 2008. Trade-offs in activity budget in an alpine ungulate: contrasting lactating and nonlactating females. Anim. Behav. 75, 217–227.

Hamilton, W.D., 1971. Geometry for the selfish herd. J. Theor. Biol. 31, 295–311.

Handegard, N.O., Boswell, K.M., Ioannou, C.C., Leblanc, S.P., Tjøstheim, D.B., Couzin, I.D., 2012. The dynamics of coordinated group hunting and collective information transfer among schooling prey. Curr. Biol. 22, 1213–1217.

Hannon, M.J., Jenkins, S.H., Crabtree, R.L., Swanson, A.K., 2006. Visibility and vigilance: behavior and population ecology of Uinta ground squirrels (*Spermophilus armatus*) in different habitats. J. Mammal. 87, 287–295.

Hanson, M.T., Coss, R.G., 2001. Age differences in arousal and vigilance in California ground squirrels (*Spermophilus beecheyi*). Dev. Psychobiol. 39, 199–206.

Hardie, S.M., Buchanan-Smith, H.M., 1997. Vigilance in single- and mixed-species groups of tamarins (*Saguinus labiatus* and *Saguinus fuscicollis*). Int. J. Primatol. 18, 219–232.

Hardy, J.W., 1961. Studies in behavior and phylogeny of certain New World jays (Garrulinae). Univ. Kansas Sci. Bull. 14, 13–149.

Hare, J.F., Campbell, K.L., Senkiw, R.W., 2014. Catch the wave: prairie dogs assess neighbours' awareness using contagious displays. Proc. R. Soc. Lond. B 281, 32153.

Harkin, E.L., van Dongen, W.F.D., Herberstein, M.E., Elgar, M.A., 2000. The influence of visual obstructions on the vigilance and escape behaviour of house sparrows, *Passer domesticus*. Aust. J. Zool. 48, 259–263.

Harley, C.M., Wagenaar, D.A., 2014. Scanning behavior in the medicinal leech *Hirudo verbana*. Plos One 9, e86120.

Hart, A., Lendrem, D.W., 1984. Vigilance and scanning patterns in birds. Anim. Behav. 32, 1216–1224.

Hart, P.J., Freed, L.A., 2005. Predator avoidance as a function of flocking in the sexually dichromatic Hawaii akepa. J. Ethol. 23, 29–33.

Harvey, P.H., Pagel, M.D., 1991. The Comparative Method in Evolutionary Biology. Oxford University Press, Oxford.

Haude, R.H., Graber, J.G., Farres, A.G., 1976. Visual observing by rhesus monkeys: some relationships with social dominance rank. Anim. Learn. Behav. 4, 163–166.

Hawlena, D., Schmitz, O.J., 2010. Physiological stress as a fundamental mechanism linking predation to ecosystem functioning. Am. Nat. 176, 537–556.

Hayes, A.R., Huntly, N.J., 2005. Effects of wind on the behavior and call transmission of pikas (*Ochotona princeps*). J. Mammal. 86, 974–981.

Heathcote, C.F., 1987. Grouping of eastern grey kangaroos in open habitats. Aust. Wildl. Res. 14, 343–348.

Hebblewhite, M., Pletscher, D., 2002. Effects of elk group size on predation by wolves. Can. J. Zool. 80, 800–809.

Heinsohn, R.G., 1987. Age-dependent vigilance in winter aggregations of cooperatively breeding white-winged choughs (*Corcorax melanorhamphos*). Behav. Ecol. Sociobiol. 20, 303–306.

Henson, P., Cooper, J.A., 1994. Nocturnal behavior of breeding trumpeter swans. Auk 111, 1013–1018.

Henson, P., Grant, T.A., 1991. The effects of human disturbance on trumpeter swan breeding behavior. Wildl. Soc. B 19, 248–257.

Herzing, D.L., Johnson, C.M., 1997. Interspecific interactions between Atlantic spotted dolphins (*Stenella frontalis*) and bottlenose dolphins (*Tursiops truncatus*) in the Bahamas, 1985–1995. Aquat. Mammals 23, 85–99.

Hews, D.K., Worthington, R.A., 2001. Fighting from the right side of the brain: left visual field preference during aggression in free-ranging male tree lizards (*Urosaurus ornatus*). Brain Behav. Evol. 58, 356–361.

Heymann, E.W., Buchanan-Smith, H.M., 2000. The behavioural ecology of mixed-species troops of Callitricine primates. Biol. Rev. 75, 169–190.

Hicklin, P.W., Gratto-Trevor, C.L., 2010. Semipalmated sandpiper. In: Poole, A. (Ed.), The Birds of North America Online. Cornell Lab of Ornithology, Ithaca.

Hill, P.S.M., 2009. How do animals use substrate-borne vibrations as an information source? Naturwissenschaften 96, 1355–1371.

Hilton, G.M., Cresswell, W., Ruxton, G.D., 1999a. Intra-flock variation in the speed of response on attack by an avian predator. Behav. Ecol. 10, 391–395.

Hilton, G.M., Ruxton, G.D., Cresswell, W., 1999b. Choice of foraging area with respect to predation risk in redshanks: the effects of weather and predator activity. Oikos 87, 295–302.

Hingee, M., Magrath, R.D., 2009. Flights of fear: a mechanical wing whistle sounds the alarm in a flocking bird. Proc. R. Soc. Lond. B 276, 4173–4179.

Hirsch, B.T., 2002. Social monitoring and vigilance behavior in brown capuchin monkeys (*Cebus apella*). Behav. Ecol. Sociobiol. 52, 458–464.

Hirsch, B.T., 2007. Costs and benefits of within-group spatial position: a feeding competition model. Q. Rev. Biol. 82, 9–27.

Hoffman, A.M., Robakiewicz, P.E., Tuttle, E.M., Rogers, L.J., 2006. Behavioural lateralization in the Australian magpie (*Gymnorhina tibicen*). Laterality 11, 110–121.

Hofmann, H.A., Beery, A.K., Blumstein, D.T., Couzin, I.D., Earley, R.L., Hayes, L.D., Hurd, P.L., Lacey, E.A., Phelps, S.M., Solomon, N.G., Taborsky, M., Young, L.J., Rubenstein, D.R., 2014. An evolutionary framework for studying mechanisms of social behavior. Trends Ecol. Evol. 29, 581–589.

Hogstad, O., 1988a. Advantages of social foraging of willow tits *Parus montanus*. Ibis 130, 275–283.

Hogstad, O., 1988b. Social rank and antipredator behaviour of willow tits *Parus montanus* in winter flocks. Ibis 130, 45–56.

Hogstad, O., 1991. The effect of social dominance on foraging by the 3-toed woodpecker *Picoides tridactylus*. Ibis 133, 271–276.

Hollen, L.I., Bell, M.B.V., Wade, H.M., Rose, R., Russell, A., Niven, F., Ridley, A.R., Radford, A.N., 2011. Ecological conditions influence sentinel decisions. Anim. Behav. 82, 1435–1441.

Hollen, L.I., Clutton-Brock, T., Manser, M.B., 2008. Ontogenetic changes in alarm-call production and usage in meerkats (*Suricata suricatta*): adaptations or constraints? Behav. Ecol. Sociobiol. 62, 821–829.

Holmes, N.D., 2007. Comparing king, gentoo, and royal penguin responses to pedestrian visitation. J. Wildl. Manage. 71, 2575–2582.

Holmes, W.G., 1984. Predation risk and foraging behavior of the hoary marmot in Alaska. Behav. Ecol. Sociobiol. 15, 263–271.

Hoogland, J.L., 1979. The effect of colony size on individual alertness of prairie dogs (Sciuridae: *Cynomys* spp.). Anim. Behav. 27, 394–407.

Hoogland, J.L., 1981. The evolution of coloniality in white-tailed and black-tailed prairie dogs (Sciuridae: *Cynomys leucurus* and *C. ludovicianus*). Ecology 62, 252–272.

Hoogland, J.L., Hale, S.L., Kirk, A.D., Sui, Y.D., 2013. Individual variation in vigilance among white-tailed prairie dogs (*Cynomys leucurus*). Southwest. Nat. 58, 279–285.

Hope, D.D., Lank, D.B., Ydenberg, R.C., 2014. Mortality-minimizing sandpipers vary stopover behavior dependent on age and geographic proximity to migrating predators. Behav. Ecol. Sociobiol. 68, 827–838.

Hopewell, L., Rossiter, R., Blower, E., Leaver, L., Goto, K., 2005. Grazing and vigilance by Soay sheep on Lundy island: influence of group size, terrain and the distribution of vegetation. Behav. Proc. 70, 186–193.

Huhta, E., Rytkönen, S., Solonen, T., 2003. Plumage brightness of prey increases predation risk: an among-species comparison. Ecology 84, 1793–1799.

Ikuta, L.A., Blumstein, D.T., 2003. Do fences protect birds from human disturbance? Biol. Conserv. 112, 447–452.

Illius, A.W., FitzGibbon, C.D., 1994. Costs of vigilance in foraging ungulates. Anim. Behav. 47, 481–484.

Inger, R., Bearhop, S., Robinson, J.A., Ruxton, G., 2006. Prey choice affects the trade-off balance between predation and starvation in an avian herbivore. Anim. Behav. 71, 1335–1341.

Inglis, I.R., Lazarus, J., 1981. Vigilance and flock size in brent geese: the edge effect. Z. Tierpsychol. 57, 193–200.

Inman, A.J., Krebs, J.R., 1987. Predation and group living. Trends Ecol. Evol. 2, 31–32.

Ioannou, C.C., Bartumeus, F., Krause, J., Ruxton, G.D., 2011. Unified effects of aggregation reveal larger prey groups take longer to find. Proc. R. Soc. Lond. B 278, 2985–2990.

Ioannou, C.C., Morrell, L.J., Ruxton, G.D., Krause, J., 2009. The effect of prey density on predators: conspicuousness and attack success are sensitive to spatial scale. Am. Nat. 173, 499–506.

Iribarren, C., Kotler, B.P., 2012. Patch use and vigilance behaviour by Nubian ibex: the role of the effectiveness of vigilance. Evol. Ecol. Res. 14, 223–234.

Jackson, A.L., Ruxton, G.D., 2006. Toward an individual-level understanding of vigilance: the role of social information. Behav. Ecol. 17, 532–538.

Jarman, P.J., Wright, S.M., 1993. Macropod studies at Wallaby Creek. IX. Exposure and responses of eastern grey kangaroos to dingoes. Aust. Wildl. Res. 20, 833–843.

Javurkova, V., Horak, D., Kreisinger, J., Klvana, P., Albrecht, T., 2011. Factors affecting sleep/vigilance behaviour in incubating mallards. Ethology 117, 345–355.

Jayakody, S., Sibbald, A.M., Gordon, I.J., Lambin, X., 2008. Red deer *Cervus elephus* vigilance behaviour differs with habitat and type of human disturbance. Wildl. Biol. 14, 81–91.

Jenkins, D.W., 1944. Territory as a result of the despotism and social organization as shown by geese. Auk 61, 30–47.

Jennings, T., Evans, S.M., 1980. Influence of position in the flock and flock size on vigilance in the starling, *Sturnus vulgaris*. Anim. Behav. 28, 634–635.

Jensen, K.H., Larsson, P., 2002. Predator evasion in *Daphnia*: the adaptive value of aggregation associated with attack abatement. Oecologia 132, 461–467.

Jiang, T.Y., Wang, X.M., Ding, Y.Z., Liu, Z.S., Wang, Z.H., 2013. Behavioral responses of blue sheep (*Pseudois nayaur*) to nonlethal human recreational disturbance. Chinese Sci. Bull. 58, 2237–2247.

Johnson, C.A., Giraldeau, L.A., Grant, J.W.A., 2001. The effect of handling time on interference among house sparrows foraging at different seed densities. Behaviour 138, 597–614.

Jones, K.A., Krebs, J.R., Whittingham, M.J., 2007. Vigilance in the third dimension: head movement not scan duration varies in response to different predator models. Anim. Behav. 74, 1181–1187.

Jones, M.E., 1998. The function of vigilance in sympatric marsupial carnivores: the eastern quoll and the Tasmanian devil. Anim. Behav. 56, 1279–1284.

Jónsson, J.E., Afton, A.D., 2009. Time budgets of snow geese *Chen caerulescens* and Ross's geese *Chen rossii* in mixed flocks: implications of body size, ambient temperature and family associations. Ibis 151, 134–144.

Jorde, D.G., Krapu, G.L., Crawford, R.D., Hay, M.A., 1984. Effects of weather on habitat selection and behavior of mallards wintering in Nebraska. Condor 86, 258–265.

Jorde, D.G., Owen, R.B., 1988. The need for nocturnal activity and energy budgets of waterfowl. In: Weller, M.W. (Ed.), Waterfowl in Winter. University of Minnesota Press, Minneapolis, pp. 169–180.

Kaby, U., Lind, J., 2003. What limits predator detection in blue tits (*Parus caeruleus*): posture, task or orientation? Behav. Ecol. Sociobiol. 54, 534–538.

Kacelnik, A., 1979. Foraging efficiency of great tits (*Parus major* L.) in relation to light intensity. Anim. Behav. 27, 237–241.

Kahlert, H., Fox, A.D., Ettrup, H., 1996. Nocturnal feeding in moulting greylag geese *Anser anser*: an anti-predator response. Ardea 84, 15–22.

Kameda, T., Tamura, R., 2007. "To eat or not to be eaten?" Collective risk-monitoring in groups. J. Exp. Soc. Psychol. 43, 168–179.

Kane, S.A., Zamani, M., 2014. Falcons pursue prey using visual motion cues: new perspectives from animal-borne cameras. J. Exp. Biol. 217, 225–234.

Kenward, R.E., 1978. Hawks and doves: factors affecting success and selection in goshawk attacks on woodpigeons. J. Anim. Ecol. 47, 449–460.

Kern, J.M., Radford, A.N., 2014. Sentinel dwarf mongooses, *Helogale parvula*, exhibit flexible decision making in relation to predation risk. Anim. Behav. 98, 185–192.

Keverne, E.B., Leonard, R.A., Scruton, D.M., Young, S.K., 1978. Visual monitoring in social groups of talapoin monkeys (*Miopithecus talapoin*). Anim. Behav. 26, 933–944.

Keys, G.C., Dugatkin, L.A., 1990. Flock size and position effects on vigilance, aggression and prey capture in the European starling. Condor 92, 151–159.

King, A.J., Cowlishaw, G., 2007. When to use social information: the advantage of large group size in individual decision making. Biol. Lett. 3, 137–139.

King, J.A., 1955. Social behavior, social organization and population dynamics in a blacktailed prairie dog town in the Black Hills of South Dakota. Contr. Lab. Vertebr. Biol. Univ. Mich. 67, 1–123.

Kingsley, M.C.S., Stirling, I., 1991. Haul-out behavior of ringed and bearded seals in relation to defense against surface predators. Can. J. Zool. 69, 1857–1861.

Kitchen, K., Lill, A., Price, M., 2011. Tolerance of human disturbance by urban magpie-larks. Aust. Field Ornithol. 28, 1–9.

Klose, S., Welbergen, J., Goldizen, A., Kalko, E., 2009. Spatio-temporal vigilance architecture of an Australian flying-fox colony. Behav. Ecol. Sociobiol. 63, 371–380.

Kluever, B.M., Breck, S.W., Howery, L.D., Krausman, P.R., Bergman, D.L., 2008. Vigilance in cattle: the influence of predation, social interactions, and environmental factors. Rangeland Ecol. Manag. 61, 321–328.

Knight, J., 2009. Making wildlife viewable: habituation and attraction. Soc. Anim. 17, 167–184.

Knight, S.K., Knight, R.L., 1986. Vigilance patterns in bald eagles feeding in groups. Auk 103, 263–272.

Koboroff, A., Kaplan, G., Rogers, L.J., 2008. Hemispheric specialization in Australian magpies (*Gymnorhina tibicen*) shown as eye preference during response to a predator. Brain Res. Bull. 76, 304–306.

Koelega, H.S., 1993. Stimulant drugs and vigilance performance: a review. Psychopharmacology 111, 1–16.

Koenker, R., 2005. Quantile Regression. Cambridge University Press, Cambridge.

Kostecke, R.M., Smith, L.M., 2003. Nocturnal behavior of American avocets in playa wetlands on the Southern High Plains of Texas, USA. Waterbirds 26, 192–195.

Kotler, B.P., Brown, J., Mukherjee, S., Berger-Tal, O., Bouskila, A., 2010. Moonlight avoidance in gerbils reveals a sophisticated interplay among time allocation, vigilance and state-dependent foraging. Proc. R. Soc. Lond. B 277, 1469–1474.

Kotrschal, K., Hemetsberger, J., Dittami, J., 1993. Food exploitation by a winter flock of greylag geese: behavioral dynamics, competition and social status. Behav. Ecol. Sociobiol. 33, 289–295.

Krakauer, D.C., 1995. Groups confuse predators by exploiting conceptual bottlenecks: a connectionist model of the confusion effect. Behav. Ecol. Sociobiol. 36, 421–429.

Krams, I., 1998. Dominance-specific vigilance in the great tit. J. Avian Biol. 29, 55–60.

Krams, I., 2001. Seeing without being seen: a removal experiment with mixed flocks of willow and crested tits *Parus montanus* and *cristatus*. Ibis 143, 476–481.

Krams, I.A., Krams, T., Cernihovics, J., 2001. Selection of foraging sites in mixed willow and crested tit flocks: rank-dependent survival strategies. Ornis Fennica 78, 1–11.

Krause, J., 1993. The effect of 'Schreckstoff' on shoaling behaviour of the minnow: a test of Hamilton's selfish herd theory. Anim. Behav. 45, 1019–1024.

Krause, J., 1994. Differential fitness returns in relation to spatial position in groups. Biol. Rev. 69, 187–206.

Krause, J., Godin, J.-G.J., 1996. Influence of prey foraging posture on flight behavior and predation risk: predators take advantage of unwary prey. Behav. Ecol. 7, 264–271.

Krause, J., Ruxton, G.D., 2002. Living in Groups. Oxford University Press, Oxford.

Krebs, J.R., 1974. Colonial nesting and social feeding as strategies for exploiting food resources in the great blue heron (*Ardea herodias*). Behaviour 51, 99–134.

Kribbs, N.B., Dinges, D., 1994. Vigilance decrement and sleepiness. In: Ogilvie, R.D., Harsh, J.R. (Eds.), Sleep Onset: Normal and Abnormal Processes. American Psychological Association, Washington, DC, US, pp. 113–125.

Kuang, F.L., Li, F.S., Liu, N., Le, F.Q., 2014. Effect of flock size and position in flock on vigilance of black-necked cranes (*Grus nigricollis*) in winter. Waterbirds 37, 94–98.

Kuijper, D.P.J., Verwijmeren, M., Churski, M., Zbyryt, A., Schmidt, K., Jedrzejewska, B., Smit, C., 2014. What cues do ungulates use to assess predation risk in dense temperate forests? Plos One 9, e84607.

Kurvers, R.H.J.M., Straates, K., Ydenberg, R.C., Wieren, S.E.v., Swierstra, P.S., Prins, H.H.T., 2014. Social information use by barnacle geese *Branta leucopsis*, an experiment revisited. Ardea 102, 173–180.

Kutsukake, N., 2006. The context and quality of social relationships affect vigilance behaviour in wild chimpanzees. Ethology 112, 581–591.

Lack, D., 1954. The Natural Regulation of Animal Numbers. Oxford University Press, Oxford.

LaGory, K.E., 1986. Habitat, group size, and the behaviour of white-tailed deer. Behaviour 98, 168–179.

Lahti, D.C., Johnson, N.A., Ajie, B.C., Otto, S.P., Hendry, A.P., Blumstein, D.T., Coss, R.G., Donohue, K., Foster, S.A., 2009. Relaxed selection in the wild. Trends Ecol. Evol. 24, 487–496.

Lajeunesse, M.J., 2009. Meta-analysis and the comparative phylogenetic method. Am. Nat. 174, 369–381.

Lamerenx, F., Chadelaud, H., Barb, B., Pépin, D., 1992. Influence of the proximity of a hiking trail on the behaviour of isards (*Rupicapra pyrenaica*) in a Pyrenean reserve. In: Spitz, F., Janeau, G., Gonzalez, G., Aulagnier, S. (Eds.), Ongulés. Institut de recherche sur les grands mammifères, Paris, pp. 605–608.

Landeau, L., Terborgh, J., 1986. Oddity and the 'confusion effect' in predation. Anim. Behav. 34, 1372–1380.

Lanham, E.J., Bull, C.M., 2004. Enhanced vigilance in groups in *Egernia stokesii*, a lizard with stable social aggregations. J. Zool. 263, 95–99.

Lank, D.B., Ydenberg, R.C., 2003. Death and danger at migratory stopovers: problems with "predation risk". J. Avian Biol. 34, 225–228.

Lark, A.M., Slade, N.A., 2008. Variation in vigilance in white-tailed deer (*Odocoileus virginianus*) in northeastern Kansas. Am. Midl. Nat. 159, 67–74.

Larsen, M.J., Sherwen, S.L., Rault, J.L., 2014. Number of nearby visitors and noise level affect vigilance in captive koalas. Appl. Anim. Behav. Sci. 154, 76–82.

Lashley, M.A., Chitwood, M.C., Biggerstaff, M.T., Morina, D.L., Moorman, C.E., DePerno, C.S., 2014. White-tailed deer vigilance: the influence of social and environmental factors. Plos One 9, e90652.

Laundré, J.W., Hernandez, L., Altendorf, K.B., 2001. Wolves, elks, and bison: reestablishing the "landscape of fear" in Yellowstone National Park, USA. Can. J. Zool. 79, 1401–1409.

Laurenson, M.K., 1994. High juvenile mortality in cheetahs (*Acinonyx jubatus*) and its consequences for maternal care. J. Zool. 234, 387–408.

Lavalli, K.L., Spanier, E., 2001. Does gregariousness function as an antipredator mechanism in the Mediterranean slipper lobster, *Scyllarides latus*. Mar. Freshwater Res. 52, 1133–1143.

Lawrence, E.S., 1985. Vigilance during 'easy' and 'difficult' foraging tasks. Anim. Behav. 33, 1373–1375.

Lazarus, J., 1972. Natural selection and the function of flocking in birds: a reply to Murton. Ibis 114, 556–558.

Lazarus, J., 1978. Vigilance, flock size and domain of danger in the white-fronted goose. Wildfowl 29, 135–145.

Lazarus, J., 1979a. The early warning function of flocking in birds: an experimental study with captive quelea. Anim. Behav. 27, 855–865.

Lazarus, J., 1979b. Flock size and behaviour in captive red-billed weaverbirds (*Quelea quelea*): implications for social facilitation and the functions of flocking. Behaviour 71, 127–145.

Lazarus, J., Inglis, I.R., 1978. The breeding behaviour of the pink-footed goose: parental care and vigilant behaviour during the feeding period. Behaviour 65, 62–88.

Lazarus, J., Inglis, I.R., 1986. Shared and unshared parental investment, parent-offspring conflict and brood size. Anim. Behav. 34, 1791–1805.

Lazarus, J., Symonds, M., 1992. Contrasting effects of protective and obstructive cover on avian vigilance. Anim. Behav. 43, 519–521.

Lea, A.J., Blumstein, D.T., 2011a. Age and sex influence marmot antipredator behavior during periods of heightened risk. Behav. Ecol. Sociobiol. 65, 1525–1533.

Lea, A.J., Blumstein, D.T., 2011b. Ontogenetic and sex differences influence alarm call responses in mammals: a meta-analysis. Ethology 117, 839–851.

Lehrer, E.W., Schooley, R.L., Whittington, J.K., 2012. Survival and antipredator behavior of woodchucks (*Marmota monax*) along an urban-agricultural gradient. Can. J. Zool. 90, 12–21.

Leighton, P.A., Horrocks, J.A., Kramer, D.L., 2010. Conservation and the scarecrow effect: can human activity benefit threatened species by displacing predators? Biol. Conserv. 143, 2156–2163.

Lemon, W.C., 1992. Fitness consequences of foraging behaviour in the zebra finch. Nature 352, 153–155.

Lendrem, D.W., 1983a. Predation risk and vigilance in the blue tit (*Parus caeruleus*). Behav. Ecol. Sociobiol. 14, 9–13.

Lendrem, D.W., 1983b. Sleeping and vigilance in birds. I. Field observations of the mallard (*Anas platyrhynchos*). Anim. Behav. 31, 532–538.

Lendrem, D.W., 1984. Flocking, feeding and predation risk: absolute and instantaneous feeding rates. Anim. Behav. 32, 298–299.

Lendrem, D.W., Stretch, D., Metcalfe, N.B., Jones, P., 1986. Scanning for predators in the purple sandpiper: a time-dependent or time-independent process? Anim. Behav. 34, 1577–1578.

Leopold, F., 1951. A study of nesting wood ducks in Iowa. Condor 53, 209–220.

Lessells, C.M., 1987. Parental investment, brood size and time budgets: behaviour of lesser snow goose families. Ardea 75, 189–203.

Leuthold, W., 1977. African Ungulates: A Comparative Review of Their Ethology and Behavioral Ecology. Springer-Verlag, New York.

Li, C., Jiang, Z., Li, L., Li, Z., Fang, H., Li, C., Beauchamp, G., 2012. Effects of reproductive status, social rank, sex and group size on vigilance patterns in Przewalski's gazelle. Plos One 7, e32607.

Li, C., Jiang, Z., Tang, S., Zeng, Y., 2007. Evidence of effects of human disturbance on alert response in Pere David's deer (*Elaphurus davidianus*). Zoo Biol. 26, 461–470.

Li, C., Yang, X., Ding, Y., Zhang, L., Fang, H., Tang, S., Jiang, Z., 2011a. Do Père David's deer lose memories of their ancestral predators? Plos One 6, e23623.

Li, C.W., Monclus, R., Maul, T.L., Jiang, Z.G., Blumstein, D.T., 2011b. Quantifying human disturbance on antipredator behavior and flush initiation distance in yellow-bellied marmots. Appl. Anim. Behav. Sci. 129, 146–152.

Li, Z., Jiang, Z., Beauchamp, G., 2009. Vigilance in Przewalski's gazelle: effects of sex, predation risk and group size. J. Zool. 277, 302–308.

Li, Z., Wang, Z., Ge, C., 2013. Time budgets of wintering red-crowned cranes: effects of habitat, age and family size. Wetlands 33, 227–232.

Li, Z.Q., Jiang, Z.G., Beauchamp, G., 2010. Nonrandom mixing between groups of Przewalski's gazelle and Tibetan gazelle. J. Mammal. 91, 674–680.

Lian, X.M., Zhang, T.Z., Cao, Y.C., Su, J.P., Thirgood, S., 2011. Road proximity and traffic flow perceived as potential predation risks: evidence from the Tibetan antelope in the Kekexili National Nature Reserve. China. Wildl. Res. 38, 141–146.

Liker, A., Barta, Z., 2002. The effects of dominance on social foraging tactic use in house sparrows. Behaviour 139, 1061–1076.

Liley, S., Creel, S., 2008. What best explains vigilance in elk: characteristics of prey, predators, or the environment? Behav. Ecol. 19, 245–254.

Lima, S.L., 1987a. Distance to cover, visual obstructions, and vigilance in house sparrows. Behaviour 102, 231–238.

Lima, S.L., 1987b. Vigilance while feeding and its relation to the risk of predation. J. Theor. Biol. 124, 303–316.

Lima, S.L., 1988. Vigilance during the initiation of daily feeding in dark-eyed juncos. Oikos 53, 12–16.

Lima, S.L., 1990. The influence of models on the interpretation of vigilance. In: Bekoff, M., Jamieson, D. (Eds.), Interpretation and Explanation in the Study of Animal Behavior: Vol. 2. Explanation, Evolution and Adaptation. Westview Press, Boulder, pp. 246–267.

Lima, S.L., 1991. Energy, predators and the behavior of feeding hummingbirds. Evol. Ecol. 5, 220–230.

Lima, S.L., 1992. Vigilance and foraging substrate: antipredatory considerations in a nonstandard environment. Behav. Ecol. Sociobiol. 30, 283–289.

Lima, S.L., 1994. On the personal benefits of anti-predatory vigilance. Anim. Behav. 48, 734–736.

Lima, S.L., 1995a. Back to the basics of anti-predatory vigilance: the group-size effect. Anim. Behav. 49, 11–20.

Lima, S.L., 1995b. Collective detection of predatory attack by social foragers: fraught with ambiguity? Anim. Behav. 50, 1097–1108.

Lima, S.L., Bednekoff, P.A., 1999a. Back to the basics of antipredatory vigilance: can nonvigilant animals detect attack? Anim. Behav. 58, 537–543.

Lima, S.L., Bednekoff, P.A., 1999b. Temporal variation in danger drives antipredator behavior: the predation risk allocation hypothesis. Am. Nat. 153, 649–659.

Lima, S.L., Dill, L.M., 1990. Behavioural decisions made under the risk of predation: a review and prospectus. Can. J. Zool. 68, 619–640.

Lima, S.L., Rattenborg, N.C., Lesku, J.A., Amlaner, C.J., 2005. Sleeping under the risk of predation. Anim. Behav. 70, 723–736.

Lima, S.L., Zollner, P.A., 1996. Anti-predatory vigilance and the limits to collective detection: visual and spatial separation between foragers. Behav. Ecol. Sociobiol. 38, 355–363.

Lima, S.L., Zollner, P.A., Bednekoff, P.A., 1999. Predation, scramble competition, and the vigilance group size effect in dark-eyed juncos (*Junco hyemalis*). Behav. Ecol. Sociobiol. 46, 110–116.

Lipetz, V.E., Bekoff, M., 1982. Group size and vigilance in pronghorns. Z. Tierpsychol. 58, 203–216.

Lippolis, G., Bisazza, A., Rogers, L.J., Vallortigara, G., 2002. Lateralisation of predator avoidance responses in three species of toads. Laterality 7, 163–183.

Lipsey, M.W., Wilson, D.B., 2001. Practical Meta-Analysis. Sage, Beverly Hills.

Loehr, J., Kovanen, M., Carey, J., Hogmander, H., Jurasz, C., Karkkainen, S., Suhonen, J., Ylonen, H., 2005. Gender- and age-class-specific reactions to human disturbance in a sexually dimorphic ungulate. Can. J. Zool. 83, 1602–1607.

Longland, W.S., Jenkins, S.H., 1987. Sex and age affect vulnerability of desert rodents to owl predation. J. Mammal. 68, 746–754.

Loonen, M.J.J.E., Bruinzel, L., Black, J.M., Drent, R.H., 1999. The benefit of large broods in barnacle geese: a study using natural and experimental manipulations. J. Anim. Ecol. 68, 753–768.

Lord, R.D., 1956. A comparative study of the eyes of some Falconiform and Passeriform birds. Am. Midl. Nat. 56, 325–344.

Loughry, W.J., 1992. Ontogeny of time allocation in black-tailed prairie dogs. Ethology 90, 206–224.

Loughry, W.J., 1993. Determinants of time allocation by adult and yearling black-tailed prairie dogs. Behaviour 124, 23–43.

Loughry, W.J., McDonough, C.M., 1989. Calling and vigilance in California ground squirrels: age, sex and seasonal differences in responses to calls. Am. Midl. Nat. 121, 312–321.

Löw, A., Lang, P.J., Smith, J.C., Bradley, M.M., 2008. Both predator and prey: emotional arousal in threat and reward. Psychol. Sci. 19, 865–873.

Lucas, E., Brodeur, J., 2001. A fox in sheep's clothing: furtive predators benefit from the communal defense of their prey. Ecology 82, 3246–3250.

Lukoschek, V., McCormick, M.I., 2002. A review of multi-species foraging associations in fishes and their ecological significance. 9th International Coral Reef Symposium. Bali. pp. 467–474.

Lung, M.A., Childress, M.J., 2007. The influence of conspecifics and predation risk on the vigilance of elk (*Cervus elaphus*) in Yellowstone National Park. Behav. Ecol. 18, 12–20.

Lynch, E., Northrup, J.M., McKenna, M.F., Anderson, C.R., Angeloni, L., Wittemyer, G., 2015. Landscape and anthropogenic features influence the use of auditory vigilance by mule deer. Behav. Ecol. 26, 75–82.

MacHutchon, A.G., Harestad, A.S., 1990. Vigilance behaviour and use of rocks by Columbian ground squirrels. Can. J. Zool. 68, 1428–1432.

Macintosh, A.J.J., Sicotte, P., 2009. Vigilance in ursine black and white colobus monkeys (*Colobus vellerosus*): an examination of the effects of conspecific threat and predation. Am. J. Primatol. 71, 919–927.

Magle, S.B., Angeloni, L.M., 2011. Effects of urbanization on the behaviour of a keystone species. Behaviour 148, 31–54.

Magurran, A.E., Higham, A., 1988. Information transfer across fish shoals under predatory threat. Ethology 78, 153–158.

Magurran, A.E., Oulton, W.J., Pitcher, T.J., 1985. Vigilant behaviour and shoal size in minnows. Z. Tierpsychol. 67, 167–178.

Makenbach, S.A., Waterman, J.M., Roth, J.D., 2013. Predator detection and dilution as benefits of associations between yellow mongooses and Cape ground squirrels. Behav. Ecol. Sociobiol. 67, 1187–1194.

Makowska, I.J., Kramer, D.L., 2007. Vigilance during food handling in grey squirrels, *Sciurus carolinensis*. Anim. Behav. 74, 153–158.

Mandalaywala, T.M., Parker, K.J., Maestripieri, D., 2014. Early experience affects the strength of vigilance for threat in *Rhesus* monkey infants. Psychol. Sci. 25, 1893–1902.

Manno, T.G., 2007. Why are Utah prairie dogs vigilant? J. Mammal. 88, 555–563.

Manor, R., Saltz, D., 2003. Impact of human nuisance disturbance on vigilance and group size of a social ungulate. Ecol. Appl. 13, 1830–1834.

Manser, M.B., 1999. Response of foraging group members to sentinel calls in suricates, *Suricata suricata*. Proc. R. Soc. Lond. B 266, 1013–1019.

Marino, A., Baldi, R., 2008. Vigilance patterns of territorial Guanacos (*Lama guanicoe*): the role of reproductive interests and predation risk. Ethology 114, 413–423.

Marler, P., 1956. Behaviour of the chaffinch *Fringilla coelebs*. Behav. Supp. 5, 1–184.

Marois, R., Ivanoff, J., 2005. Capacity limits of information processing in the brain. Trends Cogn. Sci. 9, 296–305.

Marras, S., Batty, R.S., Domenici, P., 2012. Information transfer and antipredator maneuvers in schooling herring. Adapt. Behav. 20, 44–56.

Martella, M.B., Renison, D., Navarro, J.L., 1995. Vigilance in the greater rhea: effect of vegetation height and group size. J. Field Ornithol. 66, 215–220.

Martín, B., Delgado, S., de la Cruz, A., Tirado, S., Ferrer, M., 2014. Effects of human presence on the long-term trends of migrant and resident shorebirds: evidence of local population declines. Anim. Conserv.

Martin, G.R., 2007. Visual fields and their functions in birds. J. Ornithol. 148, S547–S562.

Martín, J., Luque-Larena, J.J., Lopez, P., 2006. Collective detection in escape responses of temporary groups of Iberian green frogs. Behav. Ecol. 17, 222–226.

Martin, K., 1984. Reproductive defence priorities of male willow ptarmigan (*Lagopus lagopus*): enhancing mate survival or extending paternity options? Behav. Ecol. Sociobiol. 16, 57–63.

Maslo, B., Burger, J., Handel, S.N., 2012. Modeling foraging behavior of piping plovers to evaluate habitat restoration success. J. Wildl. Manage. 76, 181–188.

Mateo, J.M., 2007. Ecological and hormonal correlates of antipredator behavior in adult Belding's ground squirrels (*Spermophilus beldingi*). Behav. Ecol. Sociobiol. 62, 37–49.

Mateo, J.M., 2008. Inverted-U shape relationship between cortisol and learning in ground squirrels. Neurobiol. Learn. Mem. 89, 582–590.

Mateo, J.M., 2014. Development, maternal effects, and behavioral plasticity. Integr. Comp. Biol. 54, 841–849.

Mather, J.A., 2010. Vigilance and antipredator responses of Caribbean reef squid. Mar. Freshw. Behav. Physiol. 43, 357–370.

232 References

Mathews, C.G., Lesku, J.A., Lima, S.L., Amlaner, C.J., 2006. Asynchronous eye closure as an anti-predator behavior in the western fence lizard (*Sceloporus occidentalis*). Ethology 112, 286–292.

Mathis, A., Chivers, D.P., 2003. Overriding the oddity effect in mixed-species aggregations: group choice by armored and nonarmored species. Behav. Ecol. 14, 334–339.

Mathot, K.J., Giraldeau, L.-A., 2008. Increasing vulnerability to predation increases preference for the scrounger foraging tactic. Behav. Ecol. 19, 131–138.

Mathot, K.J., van den Hout, P.J., Piersma, T., Kempenaers, B., Reale, D., Dingemanse, N.J., 2011. Disentangling the roles of frequency-vs. state-dependence in generating individual differences in behavioural plasticity. Ecol. Lett. 14, 1254–1262.

Matson, T.K., Goldizen, A.W., Putland, D.A., 2005. Factors affecting the vigilance and flight behaviour of impalas. S. Afr. J. Wildl. Res. 35, 1–11.

Matthysen, E., 1999. Foraging behaviour of Nuthatches (*Sitta europaea*) in relation to the presence of mates and mixed flocks. J. Ornithol. 140, 443–451.

McCormack, J.E., Jablonski, P.G., Brown, J.L., 2007. Producer-scrounger roles and joining based on dominance in a free-living group of Mexican jays (Aphelocoma ultramarina). Behaviour 144, 967–982.

McCullough, D.R., 1993. Variation in black-tailed deer herd composition counts. J. Wildl. Manage. 57, 892–897.

McCullough, D.R., Weckerly, F.W., Garcia, P.I., Evett, R.R., 1994. Sources of inaccuracy in black-tailed deer herd composition counts. J. Wildl. Manage. 58, 319–329.

McDonald-Madden, E., Akers, L.K., Brenner, D.J., Howell, S., Patullo, B.W., Elgar, M.A., 2000. Possums in the park: efficient foraging under the risk of predation or of competition? Aust. J. Zool. 48, 155–160.

McDonough, C.M., Loughry, W.J., 1995. Influences on vigilance in nine-banded armadillos. Ethology 100, 50–60.

McGiffin, A., Lill, A., Beckman, J., Johnstone, C.P., 2013. Tolerance of human approaches by common mynas along an urban-rural gradient. Emu 113, 154–160.

McGowan, K.J., Woolfenden, G.E., 1989. A sentinel system in the Florida scrub jay. Anim. Behav. 37, 1000–1006.

McKelvey, R.W., Verbeek, N.A., 1988. Habitat use, behaviour and management of trumpeter swans, *Cygnus buccinator*, wintering at Comox, British Columbia. Can. Field Nat. 102, 434–441.

McNab, B.K., 1980. On estimating thermal conductance in endotherms. Physiol. Zool. 53, 145–156.

McNamara, J.M., Houston, A.I., 1992. Evolutionarily stable levels of vigilance as a function of group size. Anim. Behav. 43, 641–658.

McNeil, R., Drapeau, P., Goss-Custard, J.D., 1992. The occurrence and adaptive significance of nocturnal habits in waterfowl. Biol. Rev. 67, 381–419.

McNelis, N.L., Boatright-Horowitz, S.L., 1998. Social monitoring in a primate group: the relationship between visual attention and hierarchical ranks. Anim. Cogn. 1, 65–69.

McVean, A., Haddlesey, P., 1980. Vigilance schedules among house sparrows *Passer domesticus*. Ibis 122, 533–536.

Mella, V.S.A., Banks, P.B., McArthur, C., 2014a. Negotiating multiple cues of predation risk in a landscape of fear: what scares free-ranging brushtail possums? J. Zool. 294, 22–30.

Mella, V.S.A., Cooper, C.E., Davies, S., 2014b. Behavioural responses of free-ranging western grey kangaroos (*Macropus fuliginosus*) to olfactory cues of historical and recently introduced predators. Aust. Ecol. 39, 115–121.

Melzack, R., Penick, E., Beckett, A., 1959. The problem of "innate fear" of the hawk shape: an experimental study with mallard ducks. J. Comp. Physiol. Psychol. 52, 694–698.

Metcalfe, N.B., 1984. The effects of visibility on the vigilance of shorebirds: is visibility important? Anim. Behav. 32, 981–985.

Michelena, P., Deneubourg, J.-L., 2011. How group size affects vigilance dynamics and time allocation patterns: the key role of imitation and tempo. Plos One 6, e18631.

Michelena, P., Pillot, M.H., Henrion, C., Toulet, S., Boissy, A., Bon, R., 2012. Group size elicits specific physiological response in herbivores. Biol. Lett. 8, 537–539.

Milinski, M., 1977. Do all members of a swarm suffer the same predation? Z. Tierpsychol. 45, 373–388.

Miller, R.C., 1922. The significance of the gregarious habit. Ecology 3, 122–126.

Møller, A.P., Jennions, M.D., 2002. How much variance can be explained by ecologists and evolutionary biologists? Oecologia 132, 492–500.

Monaco, J.D., Rao, G., Roth, E.D., Knierim, J.J., 2014. Attentive scanning behavior drives one-trial potentiation of hippocampal place fields. Nat. Neurosci. 17, 725–731.

Monclús, R., Rödel, H.G., 2008. Different forms of vigilance in response to the presence of predators and conspecifics in a group-living mammal, the European rabbit. Ethology 114, 287–297.

Monclús, R., Rödel, H.G., 2009. Influence of different individual traits on vigilance behaviour in European rabbits. Ethology 115, 758–766.

Mooij, J.H., 1992. Behaviour and energy budget of wintering geese in the Lower Rhine area of North Rhine-Westphalia, Germany. Wildfowl 43, 121–138.

Moore, B.A., Doppler, M., Young, J.E., Fernández-Juricic, E., 2013. Interspecific differences in the visual system and scanning behavior of three forest passerines that form heterospecific flocks. J. Comp. Physiol. A Neuroethol. Sens. Neural Behav. Physiol. 199, 263–277.

Mooring, M.S., Fitzpatrick, T.A., Nishihira, T.T., Reisig, D.D., 2004. Vigilance, predation risk, and the Allee effect in desert bighorn sheep. J. Wildl. Manage. 68, 519–532.

Moreno, E., Carrascal, M., 1991. Patch residence time and vigilance in birds foraging at feeders: implications of bill shape. Ethol. Ecol. Evol. 3, 345–350.

Morgan, M.J., Godin, J.-G.J., 1985. Antipredator benefits of schooling behaviour in a cyprinodontid fish, the bandit killifish (*Fundulus diaphanus*). Z. Tierpsychol. 70, 236–246.

Morris, D., 1964. The response of animals to a restricted environment. J. Zool. 13, 99–118.

Morrison, D.A., 2014. Book review. Syst. Biol. 63, 115–117.

Mouritsen, K.N., 1992. Predator avoidance in night feeding dunlins *Calidris alpina*: a matter of concealment. Ornis Scand. 23, 195–198.

Mouritsen, K.N., 1994. Day and night feeding in dunlins *Calidris alpina*: choice of habitat, foraging technique and prey. J. Avian Biol. 25, 55–62.

Mukherjee, S., Heithaus, M.R., 2013. Dangerous prey and daring predators: a review. Biol. Rev. 88, 550–563.

Mulder, R.S., Williams, T.D., Cooke, F., 1995. Dominance, brood size and foraging behavior during brood rearing in the lesser snow goose: an experimental study. Condor 97, 99–106.

Murphy, T.G., 2007. Dishonest 'preemptive' pursuit-deterrent signal? Why the turquoise-browed motmot wags its tail before feeding nestlings. Anim. Behav. 73, 965–970.

Murton, R.K., 1968. Some predator-prey relationships in bird damage and population control. In: Murton, R.K., Wright, E.N. (Eds.), The Problems of Birds as Pests. Academic Press, London, pp. 157–169.

Murton, R.K., 1971. Why do some birds feed in flocks? Ibis 113, 534–536.

Murton, R.K., Isaacson, A.J., 1962. The functional basis of some behaviour in the wood pigeon, *Columba palumba*. Ibis 104, 503–521.

Murton, R.K., Isaacson, A.J., Westwood, N.J., 1966. The relationships between wood-pigeons and their clover food supply and the mechanisms of population control. J. Appl. Ecol. 3, 55–96.

Murton, R.K., Isaacson, A.J., Westwood, N.J., 1971. The significance of gregarious feeding behaviour and adrenal stress in a population of wood-pigeons *Columba palumbus*. J. Zool. 165, 53–84.

Nakagawa, S., Santos, E., 2012. Methodological issues and advances in biological meta-analysis. Evol. Ecol. 26, 1253–1274.

Nakayama, S., Masuda, R., Tanaka, M., 2007. Onsets of schooling behavior and social transmission in chub mackerel *Scomber japonicus*. Behav. Ecol. Sociobiol. 61, 1383–1390.

Neill, S.R.S.J., Cullen, J.M., 1974. Experiments on whether schooling by their prey affects the hunting behaviour of cephalopods and fish predators. J. Zool. 172, 549–569.

Nersesian, C.L., Banks, P.B., McArthur, C., 2012. Behavioural responses to indirect and direct predator cues by a mammalian herbivore, the common brushtail possum. Behav. Ecol. Sociobiol. 66, 47–55.

Nevin, O.T., Gilbert, B.K., 2005. Perceived risk, displacement and refuging in brown bears: positive impacts of ecotourism? Biol. Conserv. 121, 611–622.

Newberry, R.C., Estevez, I., Keeling, L.J., 2001. Group size and perching behaviour in young domestic fowl. Appl. Anim. Behav. Sci. 73, 117–129.

Norris, K.S., Dohl, T.P., 1980. The structure and function of Cetacean schools. In: Herman, L.M. (Ed.), Cetacean Behavior: Mechanisms and Functions. Wiley, New York, pp. 211–261.

Nunes, S., 2014. Maternal experience and territorial behavior in ground squirrels. J. Mammal. 95, 491–502.

Nunes, S., Muecke, E.-M., Ross, H.E., Bartholomew, P.A., 2000. Food availability affects behavior but not circulating gonadal hormones in maternal Belding's ground squirrels. Physiol. Behav. 71, 447–455.

Nystrand, M., 2007. Associating with kin affects the trade-off between energy intake and exposure to predators in a social bird species. Anim. Behav. 74, 497–506.

Ohguchi, O., 1981. Prey density and selection against oddity by three-spined sticklebacks. Z. Tierpsychol. 23, 1–79.

Öhman, A., Flykt, A., Esteves, F., 2001. Emotion drives attention: detecting the snake in the grass. J. Exp. Psychol. Gen. 130, 466–478.

Orsini, J.-P., Shaughnessy, P.D., Newsome, D., 2006. Impacts of human visitors on Australian sea lions (*Neophoca cinerea*) at Carnac Island, Western Australia: implications for tourism management. Tourism Mar. Environ. 3, 101–115.

Öst, M., Jaatinen, K., Steele, B., 2007. Aggressive females seize central positions and show increased vigilance in brood-rearing coalitions of eiders. Anim. Behav. 73, 239–247.

Öst, M., Mantila, L., Kilpi, M., 2002. Shared care provides time-budgeting advantages for female eiders. Anim. Behav. 64, 223–231.

Öst, M., Tierala, T., 2011. Synchronized vigilance while feeding in common eider brood-rearing coalitions. Behav. Ecol. 22, 378–384.

Owen, M., 1972. Some factors affecting food intake and selection in white-fronted geese. J. Anim. Ecol. 41, 79–92.

Owens, I.P.F., 2006. Where is behavioural ecology going? Trends Ecol. Evol. 21, 356–361.

Packer, C., Abrams, P., 1990. Should co-operative groups be more vigilant than selfish groups? J. Theor. Biol. 142, 341–357.

Pairah, Y.S., Prasetyo, L.B., Mustari, A.H., 2014. The time budget of Javan Deer (*Rusa timorensis*, Blainville 1822) in Panaitan Island, Ujung Kulon National Park, Banten, Indonesia. HAYATI J. Biosci. 21.

Pangle, W.M., Holekamp, K.E., 2010a. Age-related variation in threat-sensitive behavior exhibited by spotted hyenas: observational and experimental approaches. Behaviour 147, 1009–1033.

Pangle, W.M., Holekamp, K.E., 2010b. Functions of vigilance behaviour in a social carnivore, the spotted hyaena, *Crocuta crocuta*. Anim. Behav. 80, 257–267.

Pangle, W.M., Holekamp, K.E., 2010c. Lethal and nonlethal anthropogenic effects on spotted hyenas in the Masai Mara National Reserve. J. Mammal. 91, 154–164.

Pannozzo, P.L., Phillips, K.A., Haas, M.E., Mintz, E.M., 2007. Social monitoring reflects dominance relationships in a small captive group of brown capuchin monkeys (*Cebus apella*). Ethology 113, 881–888.

Papouchis, C.M., Singer, F.J., Sloan, W.B., 2001. Responses of desert bighorn sheep to increased human recreation. J. Wildl. Manage. 65, 573–582.

Pascual, J., Senar, J.C., 2013. Differential effects of predation risk and competition over vigilance variables and feeding success in Eurasian siskins (*Carduelis spinus*). Behaviour 150, 1665–1687.

Pascual, J., Senar, J.C., Domènech, J., 2014. Are the costs of site unfamiliarity compensated with vigilance? A field test in Eurasian siskins. Ethology 120, 702–714.

Pattyn, N., Neyt, X., Henderickx, D., Soetens, E., 2008. Psychophysiological investigation of vigilance decrement: boredom or cognitive fatigue? Physiol. Behav. 93, 369–378.

Pauli, J.N., Buskirk, S.W., 2007. Risk-disturbance overrides density-dependence in a hunted colonial rodent, the black-tailed prairie dog *Cynomys ludovicianus*. J. Appl. Ecol. 44, 1219–1230.

Paulus, S.L., 1984. Activity budgets of non-breeding gadwalls in Louisiana. J. Wildl. Manage. 48, 371–380.

Paulus, S.L., 1988. Time-activity budgets of mottled ducks in Louisiana in winter. J. Wildl. Manage. 52, 711–718.

Pays, O., Beauchamp, G., Carter, A.J., Goldizen, A.W., 2013. Foraging in groups allows collective predator detection in a mammal species without alarm calls. Behav. Ecol. 24, 1229–1236.

Pays, O., Blanchard, P., Valeix, M., Chamaille-Jammes, S., Duncan, P., Periquet, S., Lombard, M., Ncube, G., Tarakini, T., Makuwe, E., Fritz, H., 2012a. Detecting predators and locating competitors while foraging: an experimental study of a medium-sized herbivore in an African savanna. Oecologia 169, 419–430.

Pays, O., Blomberg, S.P., Renaud, P.C., Favreau, F.R., Jarman, P.J., 2010. How unpredictable is the individual scanning process in socially foraging mammals? Behav. Ecol. Sociobiol. 64, 443–454.

Pays, O., Dubot, A.-L., Jarman, P.J., Loisel, P., Goldizen, A.W., 2009a. Vigilance and its complex synchrony in the red-necked pademelon, *Thylogale thetis*. Behav. Ecol. 20, 22–29.

Pays, O., Ekori, A., Fritz, H., 2014. On the advantages of mixed-species groups: impalas adjust their vigilance when associated with larger prey herbivores. Ethology 120, 1207–1216.

Pays, O., Goulard, M., Blomberg, S.P., Goldizen, A.W., Sirot, E., Jarman, P.J., 2009b. The effect of social facilitation on vigilance in the eastern grey kangaroo, *Macropus giganteus*. Behav. Ecol. 20, 469–477.

Pays, O., Jarman, P., 2008. Does sex affect both individual and collective vigilance in social mammalian herbivores: the case of the eastern grey kangaroo? Behav. Ecol. Sociobiol. 62, 757–767.

Pays, O., Jarman, P.J., Loisel, P., Gerard, J.-F., 2007a. Coordination, independence or synchronization of individual vigilance in the eastern grey kangaroo? Anim. Behav. 73, 595–604.

Pays, O., Renaud, P., Loisel, P., Petit, M., Gerard, J., Jarman, P., 2007b. Prey synchronize their vigilant behaviour with other group members. Proc. R. Soc. Lond. B 274, 1287–1291.

Pays, O., Sirot, E., Fritz, H., 2012b. Collective vigilance in the greater kudu: towards a better understanding of synchronization patterns. Ethology 118, 1–9.

Penning, P.D., Parsons, A.J., Newman, J.A., Orr, R.J., Harvey, A., 1993. The effects of group size on grazing time in sheep. Appl. Anim. Behav. Sci. 37, 101–109.

Pereira, A.G., Cruz, A., Lima, S.Q., Moita, M.A., 2012. Silence resulting from the cessation of movement signals danger. Curr. Biol. 22, 627–628.

Peres, C.A., 1993. Anti-predation benefits in a mixed-species group of Amazonian tamarins. Folia Primatol. 61, 61–76.

Périquet, S., Todd-Jones, L., Valeix, M., Stapelkamp, B., Elliot, N., Wijers, M., Pays, O., Fortin, D., Madzikanda, H., Fritz, H., Macdonald, D.W., Loveridge, A.J., 2012. Influence of immediate predation risk by lions on the vigilance of prey of different body size. Behav. Ecol. 23, 970–976.

Périquet, S., Valeix, M., Loveridge, A.J., Madzikanda, H., Macdonald, D.W., Fritz, H., 2010. Individual vigilance of African herbivores while drinking: the role of immediate predation risk and context. Anim. Behav. 79, 665–671.

Peters, K.A., Otis, D.L., 2005. Using the risk-disturbance hypothesis to assess the relative effects of human disturbance and predation risk on foraging American oystercatchers. Condor 107, 716–725.

Petrie, S.A., Petrie, V., 1998. Activity budget of white-faced whistling-ducks during winter and spring in northern Kwazulu-Natal, South Africa. J. Wildl. Manage. 62, 1119–1126.

Placyk, J.S., Burghardt, G., 2011. Evolutionary persistence of chemicallyelicited ophiophagous antipredator responses in gartersnakes (*Thamnophis sirtalis*). J. Comp. Psychol. 125, 134–142.

Popp, J.W., 1988a. Scanning behaviour of finches in mixed-species groups. Condor 90, 510–512.

Popp, J.W., 1985. Changes in scanning and feeding rates with group size among American goldfinches. Bird Behav. 6, 97–98.

Popp, J.W., 1988b. Effects of food-handling time on scanning rates among American goldfinches. Auk 105, 384–385.

Porter, L.M., Garber, P.A., 2007. Niche expansion of a cryptic primate, *Callimico goeldii*, while in mixed species troops. Am. J. Primatol. 69, 1340–1353.

Portugal, S.J., Guillemain, M., 2011. Vigilance patterns of wintering Eurasian wigeon: female benefits from male low-cost behaviour. J. Ornithol. 152, 661–668.

Powell, G.V.N., 1974. Experimental analysis of the social value of flocking by starlings (*Sturnus vulgaris*) in relation to predation and foraging. Anim. Behav. 22, 501–505.

Pöysä, H., 1987. Feeding-vigilance trade-off in the teal (*Anas crecca*): effects of feeding method and predation risk. Behaviour 103, 108–121.

Pöysä, H., 1994. Group foraging, distance to cover and vigilance in the teal, *Anas crecca*. Anim. Behav. 48, 921–928.

Pravosudov, V.V., Grubb, T.C., 1995. Vigilance in the tufted titmouse varies independently with air temperature and conspecific group size. Condor 97, 1064–1067.

Pravosudov, V.V., Grubb, T.C., 1998. Body mass, ambient temperature, time of day, and vigilance in tufted titmice. Auk 115, 221–223.

Pravosudov, V.V., Grubb, T.C., 1999. Effects of dominance on vigilance in avian social groups. Auk 116, 241–246.

Price, E.O., 1984. Behavioral aspects of animal domestication. Q. Rev. Biol. 59, 1–32.

Prins, H.H.T., 1996. Ecology and Behaviour of the African Buffalo. Chapman & Hall, New York.

Prins, H.H.T., Iason, G.R., 1989. Dangerous lions and nonchalant buffalo. Behaviour 108, 262–296.

Proctor, C.J., Broom, M., Ruxton, G.D., 2006. Antipredator vigilance in birds: modelling the "edge" effect. Math. Biosci. 199, 79–96.

Pulliam, H.R., 1973. On the advantages of flocking. J. Theor. Biol. 38, 419–422.

Pulliam, H.R., Caraco, T., 1984. Living in groups: is there an optimal group size? In: Krebs, J.R., Davies, N.B. (Eds.), Behavioural Ecology. Blackwell Scientific Publications, Oxford, pp. 122–147.

Pulliam, H.R., Pyke, G.H., Caraco, T., 1982. The scanning behavior of juncos: a game-theoretical approach. J. Theor. Biol. 95, 89–103.

Putman, B.J., Clark, R.W., 2015. The fear of unseen predators: ground squirrel tail flagging in the absence of snakes signals vigilance. Behav. Ecol. 26, 185–193.

Quadros, S., Goulart, V.D.L., Passos, L., Vecci, M.A.M., Young, R.J., 2014. Zoo visitor effect on mammal behaviour: does noise matter? Appl. Anim. Behav. Sci. 156, 78–84.

Quammen, D., 1996. The Song of the Dodo: Island Biogeography in an Age of Extinctions. Scribner, New York.

Quenette, P.-Y., 1990. Functions of vigilance in mammals: a review. Acta Oecol. 11, 801–818.

Quenette, P.-Y., Gérard, J.-F., 1992. From individual to collective vigilance in wild boars (*Sus scrofa*). Can. J. Zool. 70, 1632–1635.

Quenette, P.V., Desportes, J.P., 1992. Temporal and sequential structure of vigilance behaviour of wild boars (*Sus scrofa*). J. Mammal. 73, 535–540.

Quinn, J.L., Cresswell, W., 2004. Predator hunting behaviour and prey vulnerability. J. Anim. Ecol. 73, 143–154.

Quinn, J.L., Whittingham, M.J., Butler, S.J., Cresswell, W., 2006. Noise, predation risk compensation and vigilance in the chaffinch *Fringilla coelebs*. J. Avian Biol. 37, 601–608.

Quirici, V., Castro, R.A., Oyarzun, J., Ebensperger, L.A., 2008. Female degus (*Octodon degus*) monitor their environment while foraging socially. Anim. Cogn. 11, 441–448.

Rabin, L.A., Coss, R.G., Owings, D.H., 2006. The effects of wind turbines on antipredator behavior in California ground squirrels (*Spermophilus beecheyi*). Biol. Conserv. 131, 410–420.

Radford, A.N., Bell, M.B.V., Hollen, L.I., Ridley, A.R., 2011. Singing for your supper: sentinel calling by kleptoparasites can mitigate the cost to victims. Evolution 65, 900–906.

Radford, A.N., Ridley, A.R., 2007. Individuals in foraging groups may use vocal cues when assessing their need for anti-predator vigilance. Biol. Lett. 3, 249–252.

Ramirez, J.E., Keller, G.S., 2010. Effects of landscape on behavior of black-tailed prairie dogs (*Cynomys ludovicianus*) in rural and urban habitats. Southwest. Nat. 55, 167–171.

Rand, A.L., 1954. Social feeding behavior of birds. Fieldiana Zool. 36, 1–71.

Randall, J.A., 2001. Evolution and function of drumming as communication in mammals. Am. Zool. 41, 1143–1156.

Randler, C., 2003. Vigilance in urban swan geese and their hybrids. Waterbirds 26, 257–260.

Randler, C., 2005a. Coots *Fulica atra* reduce their vigilance under increased competition. Behav. Proc. 68, 173–178.

Randler, C., 2005b. Eye preference for vigilance during feeding in coot *Fulica atra*, and geese *Anser anser* and *Anser cygnoides*. Laterality 10, 535–543.

Randler, C., 2005c. Vigilance during preening in coots *Fulica atra*. Ethology 111, 169–178.

Randler, C., 2006a. Disturbances by dog barking increase vigilance in coots *Fulica atra*. Eur. J. Wildl. Res. 52, 265–270.

Randler, C., 2006b. Is tail wagging in white wagtails, *Motacilla alba*, an honest signal of vigilance? Anim. Behav. 71, 1089–1093.

Rands, S.A., 2010. Self-improvement for team-players: the effects of individual effort on aggregated group information. Plos One 5, e11705.

Ranta, E., Peuhkuri, N., Hirvonen, H., Barnard, C.J., 1998. Producers, scroungers and the price of a free meal. Anim. Behav. 55, 737–744.

Rasa, O.A.E., 1989. Behavioural parameters of vigilance in the dwarf mongoose: social acquisition of a sex-biased role. Behaviour 110, 125–145.

Rasa, O.A.E., 1983. Dwarf mongoose and hornbill mutualism in the Taru Desert, Kenya. Behav. Ecol. Sociobiol. 12, 181–190.

Rasa, O.A.E., 1987. Vigilance behaviour in dwarf mongooses: selfish or altruistic? S. Afr. J. Sci. 83, 587–590.

Rattenborg, N.C., Amlaner, C.J., Lima, S.L., 2000. Behavioural, neurophysiological and evolutionary perspectives on unihemispheric sleep. Neurosci. Biobehav. Rev. 24, 817–842.

Rattenborg, N.C., Lima, S.L., Amlaner, C.J., 1999a. Facultative control of avian unihemispheric sleep under the risk of predation. Behav. Brain Res. 105, 163–172.

Rattenborg, N.C., Lima, S.L., Amlaner, C.J., 1999b. Half-awake to the risk of predation. Nature 397, 397–398.

Réale, D., Bousses, P., Chapuis, J.-L., 1996. Female-biased mortality induced by male sexual harassment in a feral sheep population. Can. J. Zool. 74, 1812–1818.

Reboreda, J.C., Fernandez, G.J., 1997. Sexual, seasonal and group size differences in the allocation of time between vigilance and feeding in the greater rhea, *Rhea americana*. Ethology 103, 198–207.

Rees, E.C., Bruce, J.H., White, G.T., 2005. Factors affecting the behavioural responses of whooper swans (*Cygnus c. cygnus*) to various human activities. Biol. Conserv. 121, 369–382.

Reimers, E., Lund, S., Ergon, T., 2011. Vigilance and fright behaviour in the insular Svalbard reindeer (*Rangifer tarandus platyrhynchus*). Can. J. Zool. 89, 753–764.

Repasky, R.P., 1996. Using vigilance behaviour to test whether predation promotes habitat partitioning. Ecology 77, 1880–1887.

Riddington, R., Hassall, M., Lane, S.J., Turner, P.A., Walters, R., 1996. The impact of disturbance on the behaviour and energy budgets of Brent geese *Branta b. bernicla*. Bird Study 43, 269–279.

Ridgway, S., Carder, D., Finneran, J., Keogh, M., Kamolnick, T., Todd, M., Goldblatt, A., 2006. Dolphin continuous auditory vigilance for five days. J. Exp. Biol. 209, 3621–3628.

Ridley, A.R., Nelson-Flower, M.J., Thompson, A.M., 2013. Is sentinel behaviour safe? An experimental investigation. Anim. Behav. 85, 137–142.

Ridley, A.R., Raihani, N.J., 2007. Facultative response to a kleptoparasite by the cooperatively breeding pied babbler. Behav. Ecol. 18, 324–330.

Ridley, A.R., Raihani, N.J., Bell, M.B.V., 2010. Experimental evidence that sentinel behaviour is affected by risk. Biol. Lett. 6, 445–448.

Ridley, A.R., Wiley, E.M., Thompson, A.M., 2014. The ecological benefits of interceptive eavesdropping. Funct. Ecol. 28, 197–205.

Rieucau, G., Blanchard, P., Martin, J.G.A., Favreau, F.-R., Goldizen, A.W., Pays, O., 2012. Investigating differences in vigilance tactic use within and between the sexes in eastern grey kangaroos. Plos One 7, e44801.

Rieucau, G., Giraldeau, L.-A., 2009. Group size effect caused by food competition in nutmeg mannikins (*Lonchura punctulata*). Behav. Ecol. 20, 421–425.

Rieucau, G., Martin, J.G.A., 2008. Many eyes or many ewes: vigilance tactics in female bighorn sheep *Ovis canadensis* vary according to reproductive status. Oikos 117, 501–506.

Rieucau, G., Morand-Ferron, J., Giraldeau, L.-A., 2010. Group size effect in nutmeg mannikin: between-individuals behavioral differences but same plasticity. Behav. Ecol. 21, 684–689.

Rind, M.I., Phillips, C.J.C., 1999. The effects of group size on the ingestive and social behaviour of grazing dairy cows. Anim. Sci. 68, 589–596.

Risenhoover, K.L., Bailey, J.A., 1985. Foraging ecology of bighorn sheep: implications for habitat management. J. Wildl. Manage. 49, 797–804.

Roberts, G., 1995. A real-time response of vigilance behaviour to changes in group size. Anim. Behav. 50, 1371–1374.

Roberts, G., 1996. Why individual vigilance declines as group size increases. Anim. Behav. 51, 1077–1086.

Roberts, S.C., 1988. Social influences on vigilance in rabbits. Anim. Behav. 36, 905–913.

Robinette, R.L., Ha, J.C., 2001. Social and ecological factors influencing vigilance by Northwestern crows, *Corvus caurinus*. Anim. Behav. 62, 447–452.

Roche, E.A., Brown, C.R., 2013. Among-individual variation in vigilance at the nest in colonial cliff swallows. Wilson J. Ornithol. 125, 685–695.

Rode, K.D., Farley, S.D., Robbins, C.T., 2006. Behavioral responses of brown bears mediate nutritional effects of experimentally introduced tourism. Biol. Conserv. 133, 70–80.

Rodriguez-Girones, M.A., Vasquez, R.A., 2002. Evolutionary stability of vigilance coordination among social foragers. Proc. R. Soc. Lond. B 269, 1803–1810.

Rogers, L.J., Andrew, R.J., 2002. Comparative Vertebrate Lateralization. Cambridge University Press, Cambridge.

Rogers, L.J., Zucca, P., Vallortigara, G., 2004. Advantages of having a lateralized brain. Proc. R. Soc. Lond. B 271, S420–S422.

Rolando, A., Caldoni, R., De Sanctis, A., Laiolo, P., 2001. Vigilance and neighbour distance in foraging flocks of red-billed choughs. J. Zool. 253, 225–232.

Rose, L.M., 1994. Benefits and costs of resident males to females in white-faced capuchins, *Cebus capucinus*. Am. J. Primatol. 32, 235–248.

Rose, L.M., Fedigan, L.M., 1995. Vigilance in white-faced capuchins, *Cebus capucinus*, in Costa Rica. Anim. Behav. 49, 63–70.

Rosenbaum, P.R., 1995. Quantiles in nonrandom samples and observational studies. J. Am. Stat. Assoc. 90, 1424–1431.

Ross, E.J., Deeming, D.C., 1998. Feeding and vigilance behaviour of breeding ostriches (*Struthio camelus*) in a farming environment in Britain. Br. Poult. Sci. 39, 173–177.

Roth, T.C., Cox, J.G., Lima, S.L., 2008a. Can foraging birds assess predation risk by scent? Anim. Behav. 76, 2021–2027.

Roth, T.C., Cox, J.G., Lima, S.L., 2008b. The use and transfer of information about predation risk in flocks of wintering finches. Ethology 114, 1218–1226.

Ruckstuhl, K.E., Festa-Bianchet, M., Jorgenson, J.T., 2003. Bite rates in Rocky Mountain bighorn sheep (*Ovis canadensis*): effects of season, age, sex and reproductive status. Behav. Ecol. Sociobiol. 54, 167–173.

Ruckstuhl, K.E., Neuhaus, P., 2009. Activity budgets and sociality in a monomorphic ungulate: the African oryx (*Oryx gazella*). Can. J. Zool. 87, 165–174.

Russell, C.P., 1932. Seasonal migration of mule deer. Ecol. Monogr. 2, 2–46.

Ruusila, V., Pöysä, H., 1998. Shared and unshared parental investment in the precocial goldeneye (Aves : Anatidae). Anim. Behav. 55, 307–312.

Ruxton, G.D., Roberts, G., 1999. Are vigilance sequences a consequence of intrinsic chaos or external changes? Anim. Behav. 57, 493–495.

Ruxton, G.D., Sherratt, T.N., Speed, M.P., 2004. Avoiding Attack: The Evolutionary Ecology of Crypsis, Warning Signals and Mimicry. Oxford University Press, Oxford.

Ryan, D.A., Bawden, K.M., Bermingham, K.T., Elgar, M.A., 1996. Scanning and tail-flicking in the Australian dusky moorhen (*Gallinula tenebrosa*). Auk 113, 499–501.

Salyer, J.C., Lagler, K.F., 1940. The food and habits of the American merganser during winter in Michigan, considered in relation to fish management. J. Wildl. Manage. 4, 186–219.

Sander, D., Grafman, J., Zalla, T., 2003. The human amygdala: an evolved system for relevance detection. Rev. Neurosci. 14, 303–316.

Sansom, A., Cresswell, W., Minderman, J., Lind, J., 2008. Vigilance benefits and competition costs in groups: do individual redshanks gain an overall foraging benefit? Anim. Behav. 75, 1869–1875.

Santema, P., Clutton-Brock, T., 2013. Meerkat helpers increase sentinel behaviour and bipedal vigilance in the presence of pups. Anim. Behav. 85, 655–661.

Santema, P., Teitel, Z., Manser, M., Bennett, N., Clutton-Brock, T., 2013. Effects of cortisol administration on cooperative behavior in meerkat helpers. Behav. Ecol. 24, 1122–1127.

Sastre, P., Ponce, C., Palacín, C., Martín, C.A., Alonso, J.C., 2009. Disturbances to great bustards (*Otis tarda*) in central Spain: human activities, bird responses and management implications. Eur. J. Wildl. Res. 55, 425–432.

Scannell, J., Roberts, G., Lazarus, J., 2001. Prey scan at random to evade observant predators. Proc. R. Soc. Lond. B 268, 541–547.

Schaal, A., Ropartz, P., 1985. Le comportement de surveillance chez le daim (Dama dama): effet de variables liées à l'individu, au groupe social et à l'habitat. C. R. Acad. Sci. Paris 301, 731–736.

Schaller, G.B., 1967. The Deer and the Tiger: A Study of Wildlife in India. University of Chicago Press, Chicago.

Scheel, D., 1993. Watching for lions in the grass: the usefulness of scanning and its effects during hunts. Anim. Behav. 46, 695–704.

Schlacher, T.A., Weston, M.A., Lynn, D., Connolly, R.M., 2013. Setback distances as a conservation tool in wildlife-human interactions: testing their efficacy for birds affected by vehicles on open-coast sandy beaches. Plos One 8, e71200.

Schmitt, M.H., Stears, K., Wilmers, C.C., Shrader, A.M., 2014. Determining the relative importance of dilution and detection for zebra foraging in mixed-species herds. Anim. Behav. 96, 151–158.

Schneider, K.J., 1984. Dominance, predation, and optimal foraging in white-throated sparrow flocks. Ecology 65, 1820–1827.

Schultz, R.D., Bailey, J.A., 1978. Responses of National Park elk to human activity. J. Wildl. Manage. 42, 91–100.

Schutz, C., Schulze, C.H., 2011. Scanning behaviour of foraging ruffs *Philomachus pugnax* during spring migration: is flock size all that matters? J. Ornithol. 152, 609–616.

Schwilch, R., Piersma, T., Holmgren, N.M.A., Lukas, J., 2002. Do migratory birds need a nap after a long non-stop flight? Ardea 90, 149–154.

Scriba, M.F., Rattenborg, N.C., Dreiss, A.N., Vyssotski, A.L., Roulin, A., 2014. Sleep and vigilance linked to melanism in wild barn owls. J. Evol. Biol. 27, 2057–2068.

Seddon, L.M., Nudds, T.D., 1994. The costs of raising nidifugous offspring – brood rearing by giant Canada geese (*Branta canadensis maxima*). Can. J. Zool. 72, 533–540.

Sedinger, J.S., Eichholz, M.W., Flint, P.L., 1995. Variation in brood behavior of black brant. Condor 97, 107–115.

Sedinger, J.S., Raveling, D.G., 1990. Parenting behavior of cackling Canada geese during brood rearing: division of labor within pairs. Condor 92, 174–181.

Semeniuk, C.A.D., Dill, L.M., 2006. Anti-predator benefits of mixed-species groups of cowtail stingrays (*Pastinochus sephen*) and whiprays (*Himantura uarnak*) at rest. Ethology 112, 33–43.

Severcan, C., Yamac, E., 2011. The effects of flock size and human presence on vigilance and feeding behavior in the Eurasian coot (*Fulica atra* L.) during breeding season. Acta Ethol. 14, 51–56.

Shackman, A.J., Maxwell, J.S., McMenamin, B.W., Greischar, L.L., Davidson, R.J., 2011. Stress potentiates early and attenuates late stages of visual processing. J. Neurosci. 31, 1156–1161.

Shaffery, J.P., Ball, N.J., Amlaner, C.J., Opp, M.R., 1986. Ethological sleep studies in glaucous-winged gull chicks *Larus glaucencens*. Sleep Res. 15, 61.

Shannon, G., Angeloni, L.M., Wittemyer, G., Fristrup, K.M., Crooks, K.R., 2014a. Road traffic noise modifies behaviour of a keystone species. Anim. Behav. 94, 135–141.

Shannon, G., Cordes, L.S., Hardy, A.R., Angeloni, L.M., Crooks, K.R., 2014b. Behavioral responses associated with a human-mediated predator shelter. Plos One 9, e94630.

Sharpe, L.L., Joustra, A.S., Cherry, M.I., 2010. The presence of an avian co-forager reduces vigilance in a cooperative mammal. Biol. Lett. 6, 475–477.

Shaw, J.J., Tregenza, T., Parker, G.A., Harvey, I.F., 1995. Evolutionarily stable foraging speeds in feeding scrambles: a model and an experimental test. Proc. R. Soc. Lond. B 260, 273–277.

Shepherd, S.V., Deaner, R.O., Platt, M.L., 2006. Social status gates social attention in monkeys. Curr. Biol. 16, 119–120.

Shi, J.B., Beauchamp, G., Dunbar, R.I.M., 2010. Group-size effect on vigilance and foraging in a predator-free population of feral goats (*Capra hircus*) on the Isle of Rum, NW Scotland. Ethology 116, 329–337.

Shi, J.B., Li, D.Q., Xiao, W.F., 2011. Influences of sex, group size, and spatial position on vigilance behavior of Przewalski's gazelles. Acta Theriol. 56, 73–79.

Siegfried, W.R., Underhill, L.G., 1975. Flocking as an anti-predator strategy in doves. Anim. Behav. 23, 504–508.

Sigg, H., 1980. Differentiation of female positions in Hamadryas one-male-units. Z. Tierpsychol. 53, 265–302.

Sirot, E., 2006. Social information, antipredatory vigilance and flight in bird flocks. Anim. Behav. 72, 373–382.

Sirot, E., 2012. Negotiation may lead selfish individuals to cooperate: the example of the collective vigilance game. Proc. R. Soc. Lond. B 279, 2862–2867.

Sirot, E., Pays, O., 2011. On the dynamics of predation risk perception for a vigilant forager. J. Theor. Biol. 276, 1–7.

Sirot, E., Touzalin, F., 2009. Coordination and synchronization of vigilance in groups of prey: the role of collective detection and predators' preference for stragglers. Am. Nat. 173, 47–59.

Slotow, R., Coumi, N., 2000. Vigilance in bronze mannikin groups: the contributions of predation risk and intra-group competition. Behaviour 137, 565–578.

Slotow, R., Rothstein, S.I., 1995a. Importance of dominance status and distance from cover to foraging white-crowned sparrows: an experimental analysis. Auk 112, 107–117.

Slotow, R., Rothstein, S.I., 1995b. Influence of social status, distance from cover, and group size on feeding and vigilance in white-crowned sparrows. Auk 112, 1024–1031.

Smart, S.L., Stillman, R.A., Norris, K.J., 2008. Measuring the functional responses of farmland birds: an example for a declining seed-feeding bunting. J. Anim. Ecol. 77, 687–695.

Smith, A.C., Buchanan-Smith, H.M., Surridge, A.K., Mundy, N.I., 2005. Factors affecting group spread within wild mixed-species troops of saddleback and mustached tamarins. Int. J. Primatol. 26, 337–355.

Smith, A.C., Kelez, S., Buchanan-Smith, H.M., 2004. Factors affecting vigilance within wild mixed-species troops of saddleback (*Saguinus fuscicollis*) and moustached tamarins (*S. mystax*). Behav. Ecol. Sociobiol. 56, 18–25.

Smith, S.M., Cain, J.W., 2009. Foraging efficiency and vigilance behaviour of impala: the influence of herd size and neighbour density. Afr. J. Ecol. 47, 109–118.

Sokal, R.R., Rohlf, F.J., 1995. Biometry. W.H. Freeman and Co., New York.

Sonnichsen, L., Bokje, M., Marchal, J., Hofer, H., Jedrzejewska, B., Kramer-Schadt, S., Ortmann, S., 2013. Behavioural responses of European roe deer to temporal variation in predation risk. Ethology 119, 233–243.

Sorato, E., Gullett, P.R., Griffith, S.C., Russell, A.F., 2012. Effects of predation risk on foraging behaviour and group size: adaptations in a social cooperative species. Anim. Behav. 84, 823–834.

Southwell, C., 1987. Activity pattern of the eastern grey kangaroo *Macropus giganteus*. Mammalia 51, 211–223.

Speziale, K.L., Lambertucci, S.A., Olsson, O., 2008. Disturbance from roads negatively affects Andean condor habitat use. Biol. Conserv. 141, 1765–1772.

Squires, K.A., Martin, K., Goudie, R.I., 2007. Vigilance behavior in the harlequin duck (*Histrionicus histrionicus*) during the preincubation period in labrador: are males vigilant for self or social partner? Auk 124, 241–252.

Sridhar, H., Beauchamp, G., Shanker, K., 2009. Why do birds participate in mixed-species foraging flocks? A large-scale synthesis. Anim. Behav. 78, 337–347.

Stahl, J., Tolsma, P.H., Loonen, M.J.J.E., Drent, R.H., 2001. Subordinates explore but dominants profit: resource competition in high Arctic barnacle goose flocks. Anim. Behav. 61, 257–264.

Stanford, C.B., 1998. Chimpanzee and Red Colobus: The Ecology of Predator and Prey. Harvard University Press, Cambridge.

Stankowich, T., 2003. Marginal predation methodologies and the importance of predator preferences. Anim. Behav. 66, 589–599.

Stankowich, T., 2008. Ungulate flight responses to human disturbance: a review and meta-analysis. Biol. Conserv. 141, 2159–2173.

Stankowich, T., Blumstein, D.T., 2005. Fear in animals: a meta-analysis and review of fear assessment. Proc. R. Soc. Lond. B 272, 2627–2634.

Stankowich, T., Coss, R., 2007. The re-emergence of felid camouflage with the decay of predator recognition in deer under relaxed selection. Proc. R. Soc. Lond. B 274, 175–182.

Steenbeek, R., Piek, R.C., van Buul, M., van Hooff, J.A.R.A.M., 1999. Vigilance in wild Thomas' langurs (*Presbytis thomasi*): the importance of infanticide risk. Behav. Ecol. Sociobiol. 45, 137–150.

Steer, D., Doody, J.S., 2009. Dichotomies in perceived predation risk of drinking wallabies in response to predatory crocodiles. Anim. Behav. 78, 1071–1078.

Stensland, E., Angerbjorn, A., Berggren, P., 2003. Mixed-species groups in mammals. Mammal Rev. 33, 205–223.

Stephens, D.W., Krebs, J.R., 1986. Foraging Theory. Princeton University Press, Princeton.

Stevens, M., 2013. Sensory Ecology, Behaviour, and Evolution. Oxford University Press, Oxford.

Stirrat, S.C., 2004. Activity budgets of the agile wallaby, *Macropus agilis*. Aust. J. Zool. 52, 49–64.

Stojan-Dolar, M., Heymann, E.W., 2010a. Functions of intermittent locomotion in mustached tamarins (*Saguinus mystax*). Int. J. Primatol. 31, 677–692.

Stojan-Dolar, M., Heymann, E.W., 2010b. Vigilance of mustached tamarins in single-species and mixed-species groups-the influence of group composition. Behav. Ecol. Sociobiol. 64, 325–335.

Struhsaker, T.T., 1981. Polyspecific associations among tropical rain-forest primates. Z. Tierpsychol. 57, 268–304.

Studd, M., Montgomerie, R.D., Robertson, R.J., 1983. Group size and predator surveillance in foraging house sparrows (*Passer domesticus*). Can. J. Zool. 61, 226–231.

Sullivan, K.A., 1984. The advantages of social foraging in downy woodpeckers. Anim. Behav. 32, 16–22.

Sullivan, K.A., 1985. Vigilance patterns in downy woodpeckers. Anim. Behav. 33, 328–330.

Sullivan, K.A., 1988. Ontogeny of time budgets in yellow-eyed juncos: adaptation to ecological constraints. Ecology 69, 118–124.

Sundararaj, V., McLaren, B.E., Morris, D.W., Goyal, S.P., 2012. Can rare positive interactions become common when large carnivores consume livestock? Ecology 93, 272–280.

Sundaresan, S.R., Fischhoff, I.R., Rubenstein, D.I., 2007. Male harassment influences female movements and associations in Grevy's zebra (*Equus grevyi*). Behav. Ecol. 18, 860–865.

Susskind, J.M., Lee, D.H., Cusi, A., Feiman, R., Grabski, W., Anderson, A.K., 2008. Expressing fear enhances sensory acquisition. Nat. Neurosci. 11, 843–850.

Suter, R.B., Forrest, T.G., 1994. Vigilance in the interpretation of spectral analyses. Anim. Behav. 48, 223–225.

Sutherland, W.J., 1996. From Individual Behaviour to Population Ecology. Oxford University Press, Oxford.

Tadesse, S.A., Kotler, B.P., 2011. Seasonal habitat use by Nubian ibex (*Capra nubiana*) evaluated with behavioral indicators. Isr. J. Ecol. Evol. 57, 223–246.

Tadesse, S.A., Kotler, B.P., 2012. Impact of tourism on Nubian ibex (*Capra nubiana*) revealed through assessment of behavioral indicators. Behav. Ecol. 23, 1257–1262.

Tadesse, S.A., Kotler, B.P., 2014. Effects of habitat, group-size, sex-age class and seasonal variation on the behavioural responses of the mountain nyala (*Tragelaphus buxtoni*) in Munessa, Ethiopia. J. Trop. Ecol. 30, 33–43.

Tarlow, E.M., Blumstein, D.T., 2007. Evaluating methods to quantify anthropogenic stressors on wild animals. Appl. Anim. Behav. Sci. 102, 429–451.

Tchabovsky, A.V., Krasnov, B.R., Khokhlova, I.S., Shenbrot, G.I., 2001. The effect of vegetation cover on vigilance and foraging tactics in the fat sand rat *Psammomys obesus*. J. Ethol. 19, 105–113.

Teichroeb, J.A., Sicotte, P., 2012. Cost-free vigilance during feeding in folivorous primates? Examining the effect of predation risk, scramble competition, and infanticide threat on vigilance in ursine colobus monkeys (*Colobus vellerosus*). Behav. Ecol. Sociobiol. 66, 453–466.

Tellegen, A., Lykken, D.T., Bouchard, T.J., Wilcox, K.J., Segal, N.L., Rich, S., 1988. Personality similarity in twins reared apart and together. J. Pers. Soc. Psychol. 54, 1031–1039.

Tettamanti, F., Viblanc, V.A., 2014. Influences of mating group composition on the behavioral time-budget of male and female Alpine ibex (*Capra ibex*) during the rut. Plos One 9, e86004.

Thalmann, U., Geissmann, T., 2005. New species of woolly lemur *Avahi* (Primates: Lemuriformes) in Bemaraha (Central Western Madagascar). Am. J. Primatol. 67, 371–376.

Thierry, A.-M., Brajon, S., Spée, M., Raclot, T., 2014. Differential effects of increased corticosterone on behavior at the nest and reproductive output of chick-rearing Adélie penguins. Behav. Ecol. Sociobiol. 68, 721–732.

Thompson, D.B.A., Barnard, C.J., 1983. Anti-predator responses in mixed species flocks of lapwings, golden plovers and gulls. Anim. Behav. 31, 585–593.

Thompson, V.D., 1989. Behavioral response of 12 ungulate species in captivity to the presence of humans. Zoo Biol. 8, 275–297.

Tiebout, H.M., 1996. Costs and benefits of interspecific dominance rank - are subordinates better at finding novel food locations. Anim. Behav. 51, 1375–1381.

Tilson, R., 1980. Klipspringer (*Oreotragus oreotragus*) social structure and predator avoidance in a desert canyon. Madoqua 11, 303–314.

Tinbergen, N., 1939. On the analysis of social organization among vertebrates, with special reference to birds. Am. Midl. Nat. 21, 210–234.

Tinbergen, N., 1951. The Study of Instincts. Clarendon Press, Oxford.

Tinbergen, N., 1953. The Herring Gull's World. Collins, London.

Tinbergen, N., 1963. On aims and methods of ethology. Z. Tierpsychol. 20, 410–433.

Tinkler, E., Montgomery, W.I., Elwood, R.W., 2007. Shared or unshared parental care in overwintering brent geese (*Branta bernicla hrota*). Ethology 113, 368–376.

Tisdale, V., Fernández-Juricic, E., 2009. Vigilance and predator detection vary between avian species with different visual acuity and coverage. Behav. Ecol. 20, 936–945.

Tkaczynski, P., MacLarnon, A., Ross, C., 2014. Associations between spatial position, stress and anxiety in forest baboons *Papio anubis*. Behav. Proc. 108, 1–6.

Tolman, C.W., 1965. Emotional behavior and social facilitation of feeding in domestic chicks. Anim. Behav. 13, 493–496.

Tracey, J.P., Fleming, P.J.S., 2007. Behavioural responses of feral goats (*Capra hircus*) to helicopters. Appl. Anim. Behav. Sci. 108, 114–128.

Treherne, J.E., Foster, W.A., 1981. Group transmission of predator avoidance behaviour in a marine insect: the Trafalgar effect. Anim. Behav. 29, 911–917.

Treves, A., 1998. The influence of group size and neighbors on vigilance in two species of arboreal monkeys. Behaviour 135, 1–29.

Treves, A., 1999. Within-group vigilance in red colobus and redtail monkeys. Am. J. Primatol. 48, 113–126.

Treves, A., 2000. Theory and method in studies of vigilance and aggregation. Anim. Behav. 60, 711–722.

Treves, A., Drescher, A., Ingrisano, N., 2001. Vigilance and aggregation in black howler monkeys (*Alouatta nigra*). Behav. Ecol. Sociobiol. 50, 90–95.

Treves, A., Drescher, A., Snowdon, C.T., 2003. Maternal watchfulness in black howler monkeys (*Alouatta pigra*). Ethology 109, 135–146.

Trivers, R., 1972. Parental investment and sexual selection. In: Campbell, B. (Ed.), Sexual Selection and the Descent of Man. Aldine, Chicago, pp. 136–179.

Trouilloud, W., Delisle, A., Kramer, D.L., 2004. Head raising during foraging and pausing during intermittent locomotion as components of antipredator vigilance in chipmunks. Anim. Behav. 67, 789–797.

Turchin, P., Kareiva, P., 1989. Aggregation in *Aphis varians*: an effective strategy for reducing predation risk. Ecology 70, 1008–1016.

Tyrrell, L.P., Moore, B.A., Loftis, C., Fernández-Juricic, E., 2013. Looking above the prairie: localized and upward acute vision in a native grassland bird. Sci. Rep. 3, e3231.

Uetz, G.W., Hieber, C.S., 1994. Group size and predation risk in colonial web-building spiders: analysis of attack abatement mechanisms. Behav. Ecol. 5, 326–333.

Unck, C.E., Waterman, J.M., Verburgt, L., Bateman, P.W., 2009. Quantity versus quality: how does level of predation threat affect Cape ground squirrel vigilance? Anim. Behav. 78, 625–632.

Underwood, R., 1982. Vigilance behaviour in grazing African antelopes. Behaviour 79, 81–107.

Vahl, W.K., van der Meer, J., Weissing, F.J., van Dullemen, D., Piersma, T., 2005. The mechanisms of interference competition: two experiments on foraging waders. Behav. Ecol. 16, 845–855.

Valcarcel, A., Fernández-Juricic, E., 2009. Antipredator strategies of house finches: are urban habitats safe spots from predators even when humans are around? Behav. Ecol. Sociobiol. 63, 673–685.

Valeix, M., Fritz, H., Loveridge, A.J., Davidson, Z., Hunt, J.E., Murindagomo, F., Macdonald, D.W., 2009. Does the risk of encountering lions influence African herbivore behaviour at waterholes? Behav. Ecol. Sociobiol. 63, 1483–1494.

Vallortigara, G., Rogers, L.J., Bisazza, A., Lippolis, G., Robins, A., 1998. Complementary right and left hemifield use for predatory and agonistic behaviour in toads. Neuroreport 9, 3341–3344.

Valone, T.J., Wheelbarger, A.J., 1998. The effect of heterospecifics on the group-size effect in white-crowned sparrows (*Zonotrichia leucophrys*). Bird Behav. 12, 85–90.

van der Post, D.J., de Weerd, H., Verbrugge, R., Hemelrijk, C.K., 2013. A novel mechanism for a survival advantage of vigilant individuals in groups. Am. Nat. 182, 682–688.

van der Veen, I.T., 2002. Seeing is believing: information about predators influences yellowhammer behavior. Behav. Ecol. Sociobiol. 51, 466–471.

van Honk, J., Tuiten, A., Verbaten, R., van den Hout, M., Koppeschaar, H., Thijssen, J., de Haan, E., 1999. Correlations among salivary testosterone, mood, and selective attention to threat in humans. Horm. Behav. 36, 17–24.

van Schaik, C.P., van Noordwijk, M.A., de Boer, R.J., den Tonkelaar, I., 1983. The effect of group size on time budgets and social behaviour of wild long-tailed macaques (*Macaca fascicularis*). Behav. Ecol. Sociobiol. 13, 173–181.

Ventolini, N., Ferrero, E.A., Sponza, S., Chiesa, A.D., Zucca, P., Vallortigara, G., 2005. Laterality in the wild: preferential hemifield use during predatory and sexual behaviour in the black-winged stilt. Anim. Behav. 69, 1077–1084.

Vickery, W.L., Giraldeau, L.-A., Templeton, J.J., Kramer, D.L., Chapman, C.A., 1991. Producers, scroungers and group foraging. Am. Nat. 137, 847–863.

Vilá, B.L., Cassini, M.H., 1994. Time allocation during the reproductive season in vicuñas. Ethology 97, 226–235.

Vilá, B.L., Roig, V.G., 1992. Diurnal movements, family groups and alertness of vicuña (*Vicugna vicugna*) during the late dry season in the Laguna Blanca Reserve (Catamarca, Argentina). Small Ruminant Res. 7, 289–297.

Villanueva, C., Walker, B.G., Bertellotti, M., 2014. Seasonal variation in the physiological and behavioral responses to tourist visitation in Magellanic penguins. J. Wildl. Manage. 78, 1466–1476.

Vine, I., 1971. Risk of visual detection and pursuit by a predator and the selective advantage of flocking behaviour. J. Theor. Biol. 30, 405–422.

Vine, I., 1973. Detection of prey flocks by predators. J. Theor. Biol. 40, 207–210.

Voellmy, I.K., Goncalves, I.B., Barrette, M.-F., Monfort, S.L., Manser, M.B., 2014. Mean fecal glucocorticoid metabolites are associated with vigilance, whereas immediate cortisol levels better reflect acute anti-predator responses in meerkats. Horm. Behav. 66, 759–765.

Vulinec, K., Miller, M.C., 1989. Aggregation and predator avoidance in whirligig beetles (Coleoptera: Gyrinidae). J. N. Y. Entomol. Soc. 97, 438–447.

Wahungu, G.M., Catterall, C.P., Olsen, M.F., 2001. Predator avoidance, feeding and habitat use in the red-necked pademelon, *Thylogale thetis*, at rainforest edges. Aust. J. Zool. 49, 45–58.

Waite, R.K., 1981. Local enhancement for food finding by rooks (*Corvus frugilegus*) foraging on grassland. Z. Tierpsychol. 57, 15–36.

Waite, T.A., 1987a. Dominance-specific vigilance in the tufted titmouse: effects of social context. Condor 89, 932–935.

Waite, T.A., 1987b. Vigilance in the white-breasted nuthatch: effects of dominance and sociality. Auk 104, 429–434.

Wallace, D.J., Greenberg, D.S., Sawinski, J., Rulla, S., Notaro, G., Kerr, J.N.D., 2013. Rats maintain an overhead binocular field at the expense of constant fusion. Nature 498, 65–69.

Walters, J.R., 1982. Parental behaviour in lapwings (Charadriidae) and its relationship with clutch size and mating systems. Evolution 36, 1030–1040.

Walther, F.R., 1969. Flight behaviour and avoidance of predators in Thomson's gazelle (*Gazella thomsoni* Geunther 1884). Behaviour 34, 184–221.

Wang, K., Yang, X., Zhao, J., Yu, H., Min, L., 2009. Relations of daily activity patterns to age and flock of wintering black-necked crane (*Grus nigricollis*) at Napa Lake, Shangri-La in Yunnan. Zool. Res. 30, 74–82.

Wang, Z., Li, Z.Q., Beauchamp, G., Jiang, Z.G., 2011. Flock size and human disturbance affect vigilance of endangered red-crowned cranes (*Grus japonensis*). Biol. Conserv. 144, 101–105.

Ward, C., Low, B.S., 1997. Predictors of vigilance for American crows foraging in an urban environment. Wilson Bull. 109, 481–489.

Ward, P.I., 1985. Why birds in flocks do not co-ordinate their vigilance periods. J. Theor. Biol. 114, 383–385.

Waterman, J.M., Roth, J.D., 2007. Interspecific associations of Cape ground squirrels with two mongoose species: benefit or cost? Behav. Ecol. Sociobiol. 61, 1675–1683.

Watts, D.P., 1998. A preliminary study of selective visual attention in female mountain gorillas (*Gorilla gorilla beringei*). Primates 39, 71–78.

Wawra, M., 1988. Vigilance patterns in humans. Behaviour 107, 1–17.

Webb, P.W., 1982. Avoidance responses of fathead minnow to strikes by four teleost predators. J. Comp. Physiol. A Neuroethol. Sens. Neural Behav. Physiol. 147, 371–378.

Welp, T., Rushen, J., Kramer, D.L., Festa-Bianchet, M., de Passillé, A.M., 2004. Vigilance as a measure of fear in dairy cattle. Appl. Anim. Behav. Sci. 87, 1–13.

Welty, J.C., 1968. The Life of Birds. Saunders, Philadelphia.

Westcott, P.W., 1969. Relationships among three species of jays wintering in Southeastern Arizona. Condor 71, 353–359.

Whalen, P.J., 1998. Fear, vigilance, and ambiguity: initial neuroimaging studies of the human amygdala. Curr. Dir. Psychol. Sci. 7, 177–188.

Wheeler, H.C., Hik, D.S., 2014. Giving-up densities and foraging behaviour indicate possible effects of shrub encroachment on Arctic ground squirrels. Anim. Behav. 95, 1–8.

White-Robinson, R., 1982. Inland and saltmarsh feeding of wintering brent geese in Essex. Wildfowl 33, 113–118.

White, K.S., Berger, J., 2001. Antipredator strategies of Alaskan moose: are maternal trade-offs influenced by offspring activity? Can. J. Zool. 79, 2055–2062.

White, K.S., Testa, J.W., Berger, J., 2001. Behavioral and ecologic effects of differential predation pressure on moose in Alaska. J. Mammal. 82, 422–429.

Whitfield, D.P., 1985. Raptor predation on wintering waders in southeast Scotland. Ibis 127, 544–548.

Whitfield, D.P., 2003. Redshank *Tringa totanus* flocking behaviour, distance from cover and vulnerability to sparrowhawk *Accipiter nisus* predation. J. Avian Biol. 34, 163–169.

Whittingham, M.J., Butler, S.J., Quinn, J.L., Cresswell, W., 2004. The effect of limited visibility on vigilance behaviour and speed of predator detection: implications for the conservation of granivorous passerines. Oikos 106, 377–385.

Wichman, A., Freire, R., Rogers, L.J., 2009. Light exposure during incubation and social and vigilance behaviour of domestic chicks. Laterality 14, 381–394.

Wickler, W., 1985. Coordination of vigilance in bird groups: the 'watchman's song' hypothesis. Z. Tierpsychol. 69, 250–253.

Wikenros, C., Stahlberg, S., Sand, H., 2014. Feeding under high risk of intraguild predation: vigilance patterns of two medium-sized generalist predators. J. Mammal. 95, 862–870.

Willem, V.M., 2001. Influence of weather circumstances on behaviour and hunting success of wintering long-eared owls *Asio otus*. Limosa 74, 81–86.

Williams, G.C., 1966. Adaptation and Natural Selection. Princeton University Press, Princeton.

Williams, J.J., 1903. On the use of sentinels by valley quails. Condor 5, 146–148.

Williams, T.D., Loonen, M.J.J.E., Cooke, F., 1994. Fitness consequences of parental behavior in relation to offspring number in a precocial species, the lesser snow goose. Auk 111, 563–572.

Willis, E.O., 1972. Do birds flock in Hawaii, a land without predators? Calif. Birds 3, 1–9.

Wilson, D.S., Clark, A.B., Coleman, K., Dearstyne, T., 1994. Shyness and boldness in humans and other animals. Trends Ecol. Evol. 9, 442–446.

Wilson, E.O., 1975. Sociobiology: The Modern Synthesis. Harvard University Press, Cambridge.

Wingfield, J.C., Hegner, R.E., Dufty Jr., A.M., Ball, G.F., 1990. The "Challenge Hypothesis": theoretical implications for patterns of testosterone secretion, mating systems, and breeding strategies. Am. Nat. 136, 829–846.

Winnie, J., Creel, S., 2007. Sex-specific behavioural responses of elk to spatial and temporal variation in the threat of wolf predation. Anim. Behav. 73, 215–225.

Winterbottom, J.M., 1949. Mixed bird parties in the tropics, with special reference to northern Rhodesia. Auk 66, 258–263.

Wirtz, P., Wawra, M., 1986. Vigilance and group size in *Homo sapiens*. Ethology 71, 283–286.

Wittenberger, J.F., 1978. The evolution of mating systems in grouse. Condor 80, 126–137.

Wolff, J.O., van Horn, T., 2003. Vigilance and foraging patterns of American elk during the rut in habitats with and without predators. Can. J. Zool. 81, 266–271.

Wolters, S., Zuberbuhler, K., 2003. Mixed-species associations of Diana and Campbell's monkeys: the costs and benefits of a forest phenomenon. Behaviour 140, 371–385.

Wright, J., Berg, E., De Kort, S.R., Khazin, V., Maklakov, A.A., 2001a. Cooperative sentinel behaviour in the Arabian babbler. Anim. Behav. 62, 973–979.

Wright, J., Maklakov, A.A., Khazin, V., 2001b. State-dependent sentinels: an experimental study in the Arabian babbler. Proc. R. Soc. Lond. B 268, 821–826.

Wrona, F.J., Dixon, R.W.J., 1991. Group size and predation risk: a field analysis of encounter and dilution effects. Am. Nat. 137, 186–201.

Wu, G.M., Giraldeau, L.A., 2005. Risky decisions: a test of risk sensitivity in socially foraging flocks of *Lonchura punctulata*. Behav. Ecol. 16, 8–14.

Xia, C.J., Xu, W.X., Yang, W.K., Blank, D., Qiao, J.F., Liu, W., 2011. Seasonal and sexual variation in vigilance behavior of goitered gazelle (*Gazella subgutturosa*) in western China. J. Ethol. 29, 443–451.

Yaber, M.C., Herrera, E.A., 1994. Vigilance, group size and social status in capybaras. Anim. Behav. 48, 1301–1307.

Yasué, M., Quinn, J.L., Cresswell, W., 2003. Multiple effects of weather on the starvation and predation risk trade-off in choice of feeding location in redshanks. Funct. Ecol. 17, 727–736.

Ydenberg, R.C., Dill, L.M., 1986. The economics of fleeing from predators. Adv. Study Behav. 16, 229–249.

Yorzinski, J.L., Platt, M.L., 2014. Selective attention in peacocks during predator detection. Anim. Cogn. 17, 767–777.

Zeier, H., Karten, H., 1971. The archistriatum of the pigeon: organization of afferent and efferent connections. Brain Res. 31, 313–326.

Zheng, W., Beauchamp, G., Jiang, X., Li, Z., Yang, Q., 2013. Determinants of vigilance in a reintroduced population of Père David's deer. Curr. Zool. 59, 265–270.

Subject Index

A

Alertness, 4, 14, 23, 66
Amygdala, 58
  in birds, 59
  coordinating vigilant state, 76
  indirectly monitor environment for, 58
  in rhesus monkeys, 58
Anthropogenic disturbances
  animals adaptation, 183
  factors, 174
  habituation reducing impact of, 183
  human activities, 184
  noise, 182
  urban areas, 183
  on vigilance in captive and wild animals,
    177–182
Anti-anxiolytic drugs, 56
Anti-predator behaviour, 2
  animal aggregations, 17
  dilution and collective detection, 120
  in mammals, 24
  relationship between state and vigilance, 112
Anti-predator vigilance, 21, 38, 39, 45, 174
  adaptation, 188
  with group size, 148
  in hungry animals, 106
  light regime at night influencing, 191
  modelling, 43–44
  sexual differences in, 88
  temporal organization, 45–47
    practically achieve vigilance, 46
Apprehensiveness, 3, 4
Asymptotic vigilance, 79
Attentiveness, 3, 4
Auditory vigilance, 9–10
Avahi cleesei. See Bemaraha woolly lemur
Axis deer. *See* Chital

B

Bemaraha woolly lemur, 81
Body mass
  influencing vigilance, 93
  and sentinel behaviour, 111

Body posture, 40
  detection advantage and, 41
  and predator detection, 40
Brain-imaging techniques, 79
Brain lesions, influencing vigilance, 79
Breeding, 1
  common cranes, forage in, 166
  paired, 84
  responses to single canoe, 177
  season, white-tailed ptarmigan, 166
Brown capuchin monkeys, 27
  social monitoring and vigilance, 87

C

Causation, 51
  hormonal factors, 51–56
  neural factors, 57–62
  physiological factors, 65–66
  sensory factors, 62–65
Cervus canadensis roosevelti, 81
Cheating, 47, 125
Chimpanzee males
  vigilance in, 35
Chital, 23, 25
  alarmed, 23
  anti-predator behaviour in, 23
  ultimate/functional explanation, 23
  vigilance behaviour, 23, 24
Cognitive limitations, 122
  confusion exploit, 122
Competition, 25
  contest competition for food, 26
    vigilance during search phase, 26–29
    vigilance during the exploitation phase,
      30–32
  for mates, 34–36
    female perspective, 37
    male perspective, 35–37
  intensity manipulation, studies for, 33
  non-lethal form of interaction, 25
  relationship between vigilance and, 30
    birds, 30
    mammals, 30

Printed in the United States
By Bookmasters